# Energy Use
# in the New Millennium

## *Trends in IEA Countries*

**In support of the G8 Plan of Action**

*ENERGY
INDICATORS*

11297668

## INTERNATIONAL ENERGY AGENCY

The International Energy Agency (IEA) is an autonomous body which was established in November 1974 within the framework of the Organisation for Economic Co-operation and Development (OECD) to implement an international energy programme.

It carries out a comprehensive programme of energy co-operation among twenty-six of the OECD thirty member countries. The basic aims of the IEA are:

- To maintain and improve systems for coping with oil supply disruptions.
- To promote rational energy policies in a global context through co-operative relations with non-member countries, industry and international organisations.
- To operate a permanent information system on the international oil market.
- To improve the world's energy supply and demand structure by developing alternative energy sources and increasing the efficiency of energy use.
- To assist in the integration of environmental and energy policies.

The IEA member countries are: Australia, Austria, Belgium, Canada, Czech Republic, Denmark, Finland, France, Germany, Greece, Hungary, Ireland, Italy, Japan, Republic of Korea, Luxembourg, Netherlands, New Zealand, Norway, Portugal, Spain, Sweden, Switzerland, Turkey, United Kingdom and United States. The Slovak Republic and Poland are likely to become member countries in 2007/2008. The European Commission also participates in the work of the IEA.

## ORGANISATION FOR ECONOMIC CO-OPERATION AND DEVELOPMENT

The OECD is a unique forum where the governments of thirty democracies work together to address the economic, social and environmental challenges of globalisation. The OECD is also at the forefront of efforts to understand and to help governments respond to new developments and concerns, such as corporate governance, the information economy and the challenges of an ageing population. The Organisation provides a setting where governments can compare policy experiences, seek answers to common problems, identify good practice and work to co-ordinate domestic and international policies.

The OECD member countries are: Australia, Austria, Belgium, Canada, Czech Republic, Denmark, Finland, France, Germany, Greece, Hungary, Iceland, Ireland, Italy, Japan, Republic of Korea, Luxembourg, Mexico, Netherlands, New Zealand, Norway, Poland, Portugal, Slovak Republic, Spain, Sweden, Switzerland, Turkey, United Kingdom and United States. The European Commission takes part in the work of the OECD.

**© OECD/IEA, 2007**

International Energy Agency (IEA),
Head of Communication and Information Office,
9 rue de la Fédération, 75739 Paris Cedex 15, France.

# FOREWORD

Governments around the world are increasingly aware of the urgent need to transform the way we use energy. Leaders of the Group of Eight (G8) countries acknowledged this challenge at their Gleneagles Summit in July 2005, and made a formal request for the IEA to provide advice on how to achieve a clean, clever and competitive energy future. One of the first steps in achieving the necessary changes is to better understand how we are currently using energy and the various factors that drive or restrain demand.

This publication provides to policy makers many important insights about current energy use and $CO_2$ emission patterns that will help shape priorities for future action. It uses a set of energy indicators, developed from an updated and expanded database, that describe energy use and energy-using activities across five key sectors in IEA countries: manufacturing, services, households, and passenger and freight transport. These indicators make it possible to examine how changes in energy efficiency, economic structure, income, prices and fuel mix have affected recent trends in energy use and $CO_2$ emissions.

The results show that we have much to do. A key finding is that the rate of energy efficiency improvement in IEA countries since 1990 has been less than 1% per year – much lower than in previous decades. Consequently, final energy use and $CO_2$ emissions have both increased significantly, with particularly large rises seen in the transport and service sectors. Deeper analysis reveals many significant developments of importance to policy makers. For example, appliances and air conditioning are fast approaching space heating as the most significant source of $CO_2$ emissions from households. New engine technologies and vehicle design have produced significant efficiency benefits in passenger cars. However in many countries, the benefits of introducing more efficient vehicles have been eroded by increased congestion, changes in driver behaviour and more in-car amenities. On a more encouraging note, strong efficiency improvements are having a substantial impact in manufacturing. Energy use and $CO_2$ emissions in this sector have remained almost unchanged since 1990, even though output has increased by nearly one-third.

Despite some positive developments, the overall message is clear: we are currently not on a path to a sustainable energy future. We must find new ways to accelerate the decoupling of energy use and $CO_2$ emissions from economic growth. Other work by the IEA shows that this is indeed possible: there is still substantial scope for cost-effective energy efficiency improvements in buildings, industry and transport. However, realising this potential will require strong and innovative action on the part of governments.

As part of its G8 work, the IEA has identified a number of areas in which new policies are needed immediately. These include minimum energy performance standards for appliances, strengthened energy efficiency requirements in building codes, and mandatory fuel efficiency standards for cars and small trucks. Such policies can deliver substantial energy savings, but they represent just a start. The IEA is planning further energy indicator development and analysis to guide and support the additional policies that will be needed to truly transform the way we use energy. This new work will extend beyond IEA members to include major industrialising countries.

The analysis contained in this book would not have been possible without the substantial help we received with collecting the underlying data. We are very grateful for the close collaboration of the statisticians and analysts in IEA member countries, including experts from the European Union sponsored ODYSSEE network.

This work is published under my authority as Executive Director of the IEA and does not necessarily reflect the views of the IEA member countries.

Nobuo Tanaka

*Executive Director*

# ACKNOWLEDGEMENTS

This publication was prepared by the Office of Energy Technology and R&D (ETO) in co-operation with the Energy Statistics Division (ESD) and the Office of Long-Term Co-operation and Policy Analysis (LTO).

Peter Taylor was the co-ordinator of the project and had overall responsibility for the study. The other main authors were Pierpaolo Cazzola, Michel Francoeur, Olivier Lavagne d'Ortigue, Marek Šturc, Cecilia Tam and Michael Taylor.

Neil Hirst, Director of the ETO and Noé van Hulst, Director of the LTO, provided invaluable leadership and inspiration throughout the project. Important guidance and input was provided by Robert Dixon, Head of the Energy Technology Policy Division, Jean-Yves Garnier, Head of the Energy Statistics Division, and Richard Bradley, Head of the Energy Efficiency and Environment Division.

Many other IEA colleagues have provided helpful contributions, particularly Richard Baron, Mark Ellis, Dolf Gielen, Nigel Jollands, Debra Justus, Jens Laustsen, Takao Onoda, Kanako Tanaka, Karen Treanton, Paul Waide and Julius Walker. Simone Luft helped to prepare the manuscript. The editor was Marilyn Smith.

Production assistance was provided by the IEA Communication and Information Office (CIO): Rebecca Gaghen, Muriel Custodio, Corinne Hayworth, Bertrand Sadin and Sophie Schlondorff added significantly to the material presented.

Special thanks are directed toward the following individuals for their help in collecting data: Didier Bosseboeuf, Bruno Lapillonne, Wolfgang Eichhammer and the other participants in the European Union sponsored ODYSSEE network; Colin Webb, OECD; Antigone Gikas, Eurostat; Sarah Clough and Janet Hughes, Department of Industry Tourism and Resources, Australia; Andrew Dickson and Paul Newton, Australian Bureau of Agricultural and Resource Economics; Tony Marker, Australian Greenhouse Office, Tim McIntosh and Jean-François Bilodeau, Natural Resources Canada; Robert Tromop and Harbans Aulakh, Energy Efficiency and Conservation Authority, New Zealand; Hien Dang, Ministry of Economic Development, New Zealand; Fridtjof Unander, ENOVA, Norway; Yukari Yamashita, Institute of Energy Economics, Japan; Hunter Danskin, Department for Environment, Food and Rural Affairs, United Kingdom; Julian Prime, Department for Business, Enterprise and Regulatory Reform, United Kingdom; Crawford Honeycutt, Mark Schipper and John Cymbalsky, United States Department of Energy.

We also thank the government representatives of IEA member countries, particularly the Committee on Energy Research and Technology and the Energy Efficiency Working Party, as well as others who provided valuable comments and suggestions.

# Table of *Contents*

# LIST OF FIGURES

**EXECUTIVE SUMMARY**

# LIST OF TABLES

# EXECUTIVE SUMMARY

At the Gleneagles Summit in July 2005, the Group of Eight (G8) leaders addressed the interrelated challenges of tackling climate change, promoting clean energy and achieving sustainable development. They launched the Gleneagles Plan of Action (GPOA), which identifies transforming the way we use energy as a key priority. To advance this initiative, G8 leaders asked the International Energy Agency (IEA) to play a major role in delivering elements of the GPOA, including those relating to energy efficiency in buildings, appliances, transport and industry.

As part of its response, the IEA is developing in-depth indicators to provide state-of-the-art data and analysis on energy use, efficiency developments and policy pointers. This book is a major output from the indicator work and an important contribution to the GPOA.

## Highlights

Economic growth in IEA countries in recent decades has increased personal wealth and created many new opportunities for individuals. People travel more, and own more and larger cars. They have more spacious and comfortable homes, with a greater number and variety of appliances. They enjoy a greater range and higher quality of shops, leisure facilities, schools, hospitals, and other services. IEA economies are steadily growing.

**Figure ES.1** ▶ *Impact of Energy Efficiency Improvements on Final Energy Use, IEA11*

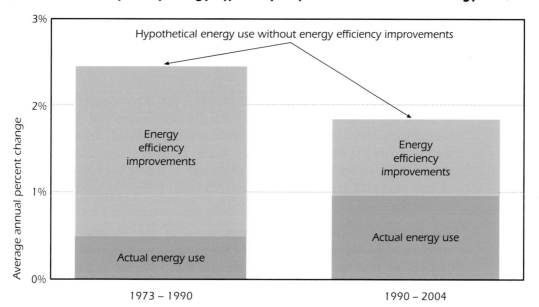

This is all good news. But it has also created greater demand for the services that energy provides (*e.g.* heating, lighting, transportation). Increased service demand need not have led to a rise in actual energy use, provided that improvements in energy efficiency kept pace. However, this was not the case. In fact, since 1990 the rate of improvement in energy efficiency has been about half of what it was in previous decades. Had the earlier rate been sustained, there would have been almost no increase in energy consumption in the IEA. Instead, final energy use increased by 14% between 1990 and 2004. This increased energy use fed directly into the level of $CO_2$ emissions, which also rose by 14%. By 2004, on average, each IEA citizen's energy consumption at home and for travel produced more than five tonnes of $CO_2$ emissions per year. Commercial and industrial activities, including the transport of goods, generated an additional six tonnes per person per year.

These findings confirm the conclusions of previous IEA analyses – that the changes caused by the oil price shocks in the 1970s and the resulting energy policies did considerably more to control growth in energy demand and reduce $CO_2$ emissions than the energy efficiency and climate policies implemented since the 1990s. Projections published by the IEA in *Energy Technology Perspectives 2006* show that the recent rate of improvement in energy efficiency will need to at least double for a realistic chance of a more sustainable energy future.

# Data

Detailed, timely and accurate information is vital to monitor the impacts of existing energy policies and to develop the new policies that will be needed. Since the last IEA indicators analysis, published in 2004, six more IEA member countries have been added to the database. This brings to 20 the total included in at least part of the current analysis. However, building complete and reliable databases requires time, expertise and resources. Even for these countries, data are not always available for all sectors and a great deal of work remains to be done in terms of improving data quality and comparability.

# Analysis

In order to change the current patterns of energy use, it is necessary to understand, in detail, the trends in energy efficiency and the other factors that influence energy consumption. This study provides an analysis of energy use and $CO_2$ emissions in IEA member countries[1] from 1990 to 2004. It uses a powerful analytical tool – energy indicators – to separate out the effects of changes in activity levels, structure (the mix of activities in the economy) and energy intensities (which are used as a proxy for energy efficiency).

The demand for energy services, which combines the impacts of activity and structure, in IEA countries increased by 1.8% per year over the period 1990 to 2004. This was less than the annual growth rate of GDP at 2.3%, due to the fact that some energy-using activities grew more slowly than the economy as a whole. All parts of

1. Data are available for the period 1990 – 2004 for all sectors for a group of 14 IEA countries: Austria, Canada, Denmark, Finland, France, Germany, Italy, Japan, the Netherlands, New Zealand, Norway, Sweden, the United Kingdom and the United States. A further six countries – Australia, Belgium, Greece, Ireland, Portugal and Spain – are included in the analysis of some sectors. In the analysis of long-term trends from 1973, 11 countries are included. These are Australia, Denmark, Finland, France, Germany, Italy, Japan, Norway, Sweden, the United Kingdom and the United States.

the economy contributed to this increased demand for energy services. The highest rates of growth were in the service and domestic passenger and freight transport sectors, followed by manufacturing and households. Even in the wealthiest countries, consumers have shown a robust and sustained desire to enhance their lifestyles in ways that require more energy services. This is borne out by detailed analysis of the key drivers of energy demand, such as ownership and use of vehicles, housing size and occupancy, and floor space and electricity demand in the service sector.

About half of the increased demand for energy services was met through increased energy use, and the other half through improvements in energy efficiency, which averaged 0.9% per year between 1990 and 2004. These improvements led to energy and $CO_2$ savings of 14% in 2004 (16 EJ [370 Mtoe] and 1.2 Gt $CO_2$). This is approximately equivalent to the annual final energy consumption and $CO_2$ emissions of Japan and translates into fuel and electricity cost savings of at least USD 170 billion in 2004. This illustrates the critical importance of energy efficiency in shaping energy use and $CO_2$ patterns. However, the efficiency gains were much lower than in previous decades; energy efficiency improvements averaged 2% per year between 1973 and 1990.

**Figure ES.2** ▶ *Energy Savings from Improvements in Energy Efficiency since 1990, IEA14*

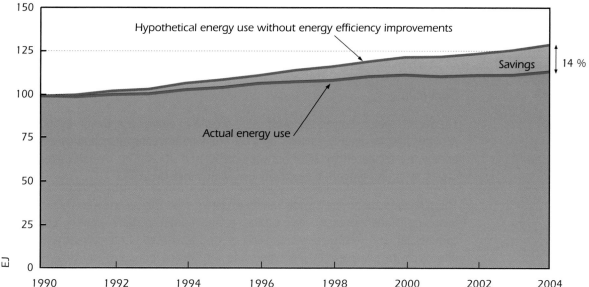

## Manufacturing

Of all sectors analysed, manufacturing industry achieved the largest energy savings from improved energy efficiency, totalling 21% in 2004. Improvements in energy efficiency, coupled with structural changes within the sector, were almost sufficient to offset the effect of growing output on energy demand. Actual energy use grew by 3%; $CO_2$ emissions increased by only 1%. The low rise in $CO_2$ emissions was helped by a shift in energy use from coal and oil to natural gas.

Significant differences in the energy intensity of manufacturing among IEA countries are explained, in part, by different industrial structures. Normalising for these differences has a large impact. For instance, most of the exceptionally high energy intensity of the Australian manufacturing sector is explained by its structure: raw materials production accounts for nearly 50% of manufacturing output. A much more detailed analysis of global industrial energy efficiency is contained in the IEA publication, *Tracking Industrial Energy Efficiency and $CO_2$ Emissions*, published in June 2007.

## Households

In the household sector, energy savings of 11% from efficiency improvements were not nearly enough to counteract growth in the demand for energy services. Household energy use, adjusted for yearly variations in climate, increased by 14%, driven largely by a spectacular 48% growth in electricity use in appliances. Space heating remains, by far, the largest user of energy in this sector. However, because of the high $CO_2$ intensity of electricity in most countries, appliances (including air conditioning) are rapidly catching up as the main cause of $CO_2$ emissions. Growth was particularly strong in the use of a wide range of small appliances. Some progress is being made in space heating; the level of heat demand per unit of floor area fell by 16% between 1990 and 2004.

Even allowing for climate variations among countries, the highest per capita household energy use (in the United States) was 2.5 times greater than the lowest (in New Zealand). This variation reflects a combination of factors including dwelling sizes, building design and comfort levels.

## Services

The service sector showed the most rapid growth in energy use, driven by increasing activity. Despite energy savings of 17% from improvements in energy efficiency, energy use increased by 26% between 1990 and 2004. This overall increase was driven by strong growth in electricity use, which rose by 50%.

The service sector covers a wide variety of activities in the private and public sectors including trade, finance, real estate, public administration, health, education, commercial and leisure services. Lack of disaggregated data currently makes it impossible to analyse precisely where the very strong pressures for greater energy demand arise.

## Passenger Transport

Energy efficiency improvements in domestic passenger transport resulted in energy savings of 7% by 2004. The vast majority (88%) of energy use in passenger transport is in cars. There have been improvements in vehicle and engine technologies but some of these gains have been offset by consumer preferences for larger and heavier vehicles and by increasing congestion in some countries. Combined with an increasing demand for passenger travel, these trends have driven up energy use in passenger transport (excluding international aviation) by 25% between 1990 and 2004.

The growth in $CO_2$ emissions per capita from passenger transport was strikingly uneven from country to country. In some countries, they remained relatively stable. Emissions actually declined in Finland, Germany and the United Kingdom, reflecting limited growth in already high vehicle ownership, limited increases in distances travelled, and improved fleet efficiency (including a shift from gasoline towards diesel). In contrast, Japan had one of the most rapid rates of $CO_2$ emissions growth. This reflects its starting point as having one of the lowest per capita emissions due to a historically high share of rail transport and a subsequent shift towards greater travel by car.

### Freight Transport

Increased levels of freight haulage have led to a 24% rise in final energy use, despite 9% savings from energy efficiency gains. Overall, the energy intensity of freight transport has declined. Improvements in the efficiency of trucking, mainly due to higher load factors and higher vehicle efficiencies, have been sufficient to offset the impacts of increases in the share of trucking, which is more energy intensive than rail or ships.

There are large differences in the level of freight $CO_2$ emissions per unit of GDP. Australia and Canada have the highest $CO_2$ emissions per unit of GDP, reflecting the importance of raw material production and the large distances over which it must be carried. However, these countries, along with the United States, have the lowest energy use per tonne-kilometre carried.

# Conclusions

Since 1990, improvements in energy efficiency have continued to play a key role in shaping energy use and $CO_2$ emissions patterns in IEA countries. By 2004, these improvements had led to an annual energy saving of 16 EJ, which is equivalent to 1.2 Gt of avoided $CO_2$ emissions and an estimated USD 170 billion of energy cost savings.

However, it is clear that more could be done; energy efficiency gains have been relatively modest since 1990 and significantly lower than in previous decades. Some small encouragement can be found in the fact that the rate of improvement seems to be increasing slightly in the last few years of the period analysed. It must be acknowledged, however, that energy efficiency measures need time to take effect. Thus, the results contained in this report may not fully reflect the impact of many policies recently initiated.

Nevertheless, there is a clear need to substantially increase the rate of energy efficiency improvement in order to tackle climate change and move towards a more secure and sustainable energy future. This is indeed possible; there is still significant scope in IEA countries for adopting more cost-effective energy-efficient technologies in buildings, industry and transport.

This book highlights a number of striking trends in the continuing growth of energy demand – trends that policy makers should take into account when developing and implementing energy-efficiency and carbon-saving strategies. Two in particular are worth mentioning here. First is the rapid increase in electricity consumption, driven by higher electricity demand from the household and service sectors. Second is the continuing growth of passenger and freight transport activity and the low rate of overall improvement in energy efficiency within these sectors, despite better vehicle and engine technologies. Governments need to establish and maintain a comprehensive framework to monitor energy consumption trends at an end-use level and should support urgent work to address the gaps in available statistical data.

Detailed analysis shows that increased demand for energy services remains deeply rooted in the lifestyle ambitions of consumers. Very strong action is needed across all sectors if this rising demand is to be counteracted by gains in energy efficiency. Governments must act now to develop and implement the necessary mix of market- and regulatory-based policies, including stringent norms and standards. This should be complemented by efforts to drive down the $CO_2$ intensity of electricity production by moving towards a cleaner technology mix.

# INTRODUCTION

This publication is part of the response from the International Energy Agency (IEA) to a request from the Group of Eight (G8) leaders to support the Gleneagles Plan of Action (GPOA), launched in July 2005. The GPOA addresses the global challenges of tackling climate change, promoting clean energy and achieving sustainable development. In particular, it identifies improvements to energy efficiency as having benefits for both economic growth and the environment. The IEA is therefore developing in-depth indicators to provide state-of-the-art data and analysis on energy use and efficiency developments, which can better inform policy-making.

This major output from the IEA work on energy indicators extends the analysis presented in *Oil Crises and Climate Challenges: 30 Years of Energy Use in IEA Countries* (IEA, 2004). It draws on detailed end-use information about the patterns of energy consumption in the manufacturing, household, service and transport sectors of 20 IEA countries over the period from 1990 to 2004. This information, coupled with economic and demographic data, is used to identify the factors behind increasing energy use and those that restrain it. The indicators include measures of activity (such as manufacturing output or volume of freight haulage), measures of developments in structure (such as changes in manufacturing output mix or modal shares in transport), and measures of energy intensity (defined as energy use per unit of activity).

Using the most detailed data available on a consistent basis across IEA countries, a decomposition approach is applied to separate and quantify the impacts of changes in activity, structure and energy intensities on final energy use in each sector. The results of the sector decompositions are then aggregated to analyse economy-wide trends. Going one step further, the decomposition approach is also used to analyse changes in $CO_2$ emissions. In this case, two additional components are investigated: fuel mix (measured as the relative shares of each fuel in the total) and carbon intensity (which, in most sectors, is related to changes in $CO_2$ emissions per unit of electricity and heat production).

This book provides an integrated analysis of how energy efficiency in all end-use sectors has affected recent developments in energy use and $CO_2$ emissions. It also examines other factors such as economic structure, income, lifestyle, prices and fuel mix, as well as the impacts of policy measures. When considering the results, readers should keep in mind the following two points. First, the indicator results are influenced by the IEA's choice of decomposition method, as well as by the measures selected for activity, structure and intensity. Some of the countries featured have undertaken their own indicators analyses and may have used other approaches. This could lead to a difference in the results obtained. Second, trends in energy intensity are only a proxy for changes in the technical level of energy efficiency. For example, the results show that some countries have experienced only a modest decline in the energy consumed per vehicle-kilometre travelled by passenger cars. However, in many cases, individual car models have become substantially more efficient. Important improvements in engine and vehicle technology have been partially offset by an increase in the share of larger, heavier and more powerful cars.

In order to assist the reader, all chapters follow a common format, starting with a short description of the sector's main patterns in energy use and $CO_2$ emissions over the period from 1990 to 2004. This is followed by a more detailed examination of the drivers of energy use, trends in aggregate energy intensity, the decomposition of these trends, an estimation of energy and $CO_2$ savings and, finally, a summary. Chapter 2 highlights the overall trends, examining energy use and $CO_2$ emissions across all end-use sectors. Each of the subsequent five chapters explores one end-use sector in more detail: manufacturing, households, services, passenger transport and freight transport. Chapter 8 summarises key conclusions and discusses the policy implications. The annexes provide a more thorough explanation of the methodology and data sources, a discussion of longer-term trends and a description of current indicators activities in IEA countries.

# OVERALL TRENDS

## Scope

This chapter examines energy use by final consumers in all main end-use sectors: manufacturing, households and services, as well as both passenger and freight transport (excluding international aviation). It does not include analysis of the fuels used in the transformation sector for the production of electricity and heat. $CO_2$ emissions that result from final energy use are also examined, including indirect emissions from the use of electricity and heat.

## *Highlights*

Between 1990 and 2004, the overall improvement in energy efficiency in a group of 14 IEA countries was 0.9% per year. Without the energy savings resulting from these improvements, total final energy consumption in the IEA14 would have been 14% higher in 2004 (Figure 2.1). This represents an annual energy saving of 16 EJ in 2004, and 1.2 Gt of avoided $CO_2$ emissions. This is approximately equivalent to the annual final energy consumption and $CO_2$ emissions of Japan, and represents an estimated USD 170 billion of energy cost savings in 2004 alone.

However, the rate of energy efficiency improvement was much lower than in previous decades. It was not nearly sufficient to offset the growth in energy use and $CO_2$ emissions, both of which increased by 14% between 1990 and 2004. For the most part, this growth was driven by increased energy use in buildings and in transport. In its publication *Energy Technology Perspectives 2006*, the IEA projects that the recent rate of improvement in energy efficiency will need to at least double in order to create a realistic chance of achieving a more sustainable energy future.

**Figure 2.1** ▶ *Economy-wide Energy Savings from Improvements in Energy Efficiency, IEA14*

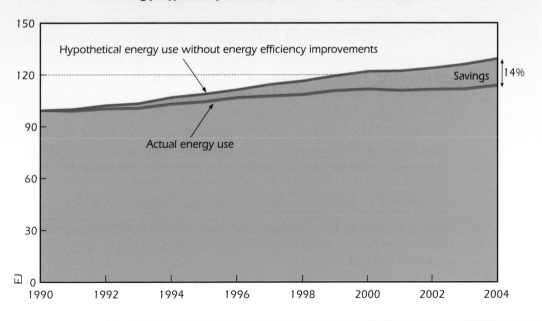

# Overview of Trends

The overview of economy-wide trends during the period 1990 to 2004 covers 14 IEA countries for which data are available for all sectors[2] (Figure 2.2). Over this time, economic activity (as measured by gross domestic product or GDP) increased by 38%. Total final energy use increased by only 14%. This partial decoupling of energy use from economic growth resulted in a 17% decrease in final energy intensity (measured as total final energy use per unit of GDP). $CO_2$ emissions, including those from the production of electricity and heat, have risen in line with energy use, also showing an increase of 14%.

**Figure 2.2** ▶ *Overview of Key Economy-wide Trends, IEA14*

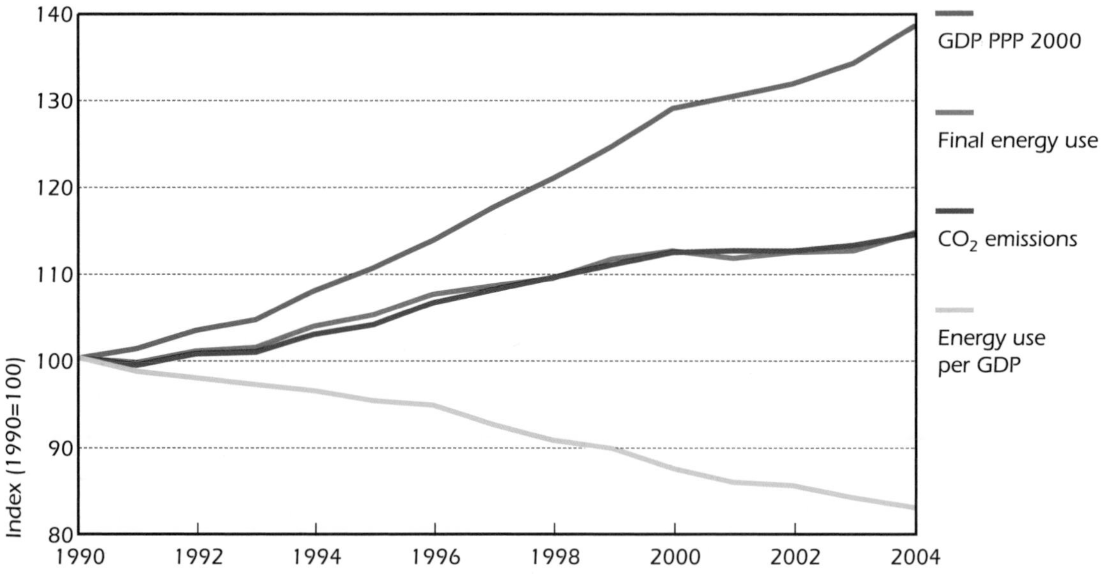

# Trends in Energy Use and CO$_2$ Emissions

### *Energy Consumption by Sector*

Since 1990, domestic passenger transport has been the sector with the fastest growth in final energy consumption in the IEA14 (Figure 2.3); in 2004, it reached a share of 26%. The increase in energy use for transporting freight was only slightly less than that for passengers; by 2004, its share rose to 11%. Energy use in buildings also increased significantly in both households and services; in 2004, these sectors consumed, respectively, 22% and 14% of total final energy use. In contrast, energy consumption in manufacturing showed only a small rise. Historically, this sector has been the largest final energy user, but its share has fallen since 1990. In 2004, it represented 26% of the total, the same share as passenger transport.

When re-grouped by main end-uses, the shares of final energy consumption in the IEA14 are as follows: passenger and freight transport – 37%; buildings – 36%; and manufacturing – 26%.

---

2. The 14 IEA countries included in the analysis of the overall trends are Austria, Canada, Denmark, Finland, France, Germany, Italy, Japan, the Netherlands, New Zealand, Norway, Sweden, the United Kingdom and the United States. These countries account for 85% of total final energy use in all IEA countries.

**Figure 2.3** ▶ *Total Final Energy Consumption by Sector, IEA14*

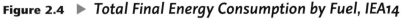

Manufacturing

Passenger transport

Households*

Services

Freight transport

Other**

*Corrected for yearly climate variations.
**Other is construction and agriculture.

## Energy Consumption by Fuel

Total final energy consumption (TFC) in the IEA14 is dominated by oil, with a share of 47% in 2004 – approximately the same as in 1990. With a share of 22%, electricity has overtaken natural gas as the second most important energy commodity in the final energy mix (Figure 2.4). In contrast, coal consumption has declined and now represents only 5% of total energy use by final users. Renewable energy use (mostly biomass) has grown, but its share remains 4% of final energy consumption.

**Figure 2.4** ▶ *Total Final Energy Consumption by Fuel, IEA14*

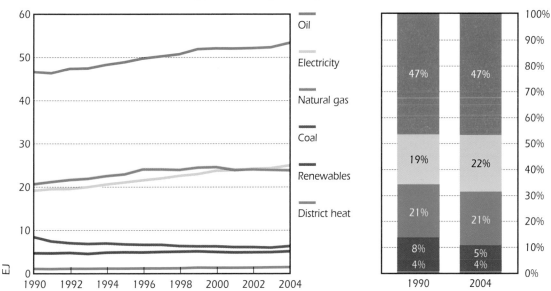

Oil

Electricity

Natural gas

Coal

Renewables

District heat

Oil use for both passenger and freight transport increased strongly between 1990 and 2004, driving the overall rise in oil demand (Figure 2.5). As a result, the transport sector now accounts for over 75% of oil consumption in end-use sectors. Conversely, oil use fell in all stationary applications (manufacturing, households and services).

**Figure 2.5** ▶ *Changes in Oil Demand and Oil Shares by Sector, IEA14*

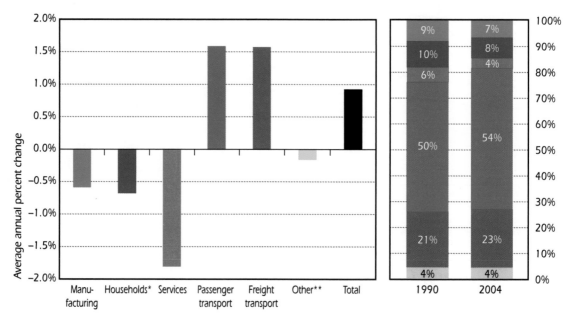

*Corrected for yearly climate variations.
**Other is construction and agriculture.

In contrast to oil, electricity use increased rapidly in all stationary sectors, with the strongest growth in households and services (Figure 2.6). Residential electricity demand is largely being driven by increased ownership and use of electric appliances. In particular, there has been a substantial increase in consumption from a wide range of smaller appliances such as home entertainment and kitchen equipment. In the service sector, much of the strong growth has been from air conditioning and lighting, and from various kinds of office and information technology equipment. Growth of electricity use in manufacturing has been less rapid. Electricity use in the remaining sectors is very small, accounting collectively for less than 3% of total electricity use.

## CO$_2$ Emissions

Rising energy use has led to increased CO$_2$ emissions in most sectors (Figure 2.7).[3] Of the major sectors, manufacturing was the only one in which CO$_2$ emissions fell between 1990 and 2004. Although manufacturing energy use grew slightly, this was more than offset by the shift towards a fuel mix that was less carbon-intensive, reflecting a decline in the shares of coal and oil and an increase in natural gas. Despite the emission reductions, manufacturing remained responsible for the highest share (24%) of total IEA14 emissions in 2004.

---

3. Throughout this publication, emissions from electricity and heat are allocated to end-use sectors (see Box 2.1).

**Figure 2.6** ▶ *Changes in Electricity Demand and Electricity Shares by Sector, IEA14*

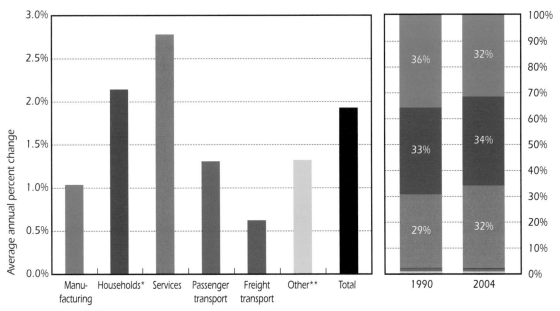

*Corrected for yearly climate variations.
**Other is construction and agriculture.

$CO_2$ emissions from the household sector increased in line with total emissions, maintaining their share at 23% in 2004. A somewhat stronger trend was seen in the service sector: changes in the fuel mix coupled with a rapid increase in final energy use means services were responsible for 17% of total $CO_2$ emissions in 2004.

Significantly higher energy use for transport also increased the share of these sectors in total $CO_2$ emissions, to 22% for passenger travel and 10% for freight haulage.

**Figure 2.7** ▶ *Changes in $CO_2$ Emissions and Emission Shares by Sector, IEA14*

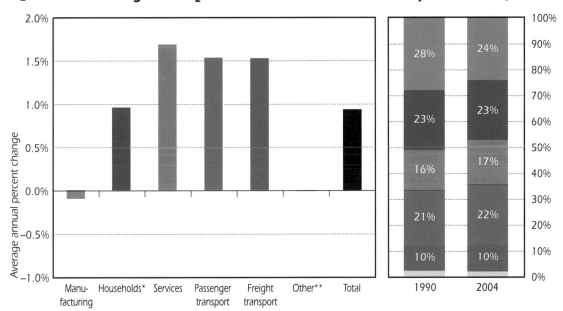

*Corrected for yearly climate variations.
**Other is construction and agriculture.

Box 2.1

# $CO_2$ Emissions from Electricity and Heat Production

In this publication, $CO_2$ emissions from final energy use – in the economy as a whole and in each sector – include indirect emissions from the use of electricity and heat. Emissions are allocated to the sector in which the electricity and heat are used, based on average yearly $CO_2$ emission coefficients for electricity and district heat, respectively. Thus, the overall and sectoral trends in $CO_2$ emissions reflect both changes in the final energy mix and developments in the $CO_2$ intensity of electricity and heat production.

In many IEA countries, the production of electricity and heat is largely based on the use of fossil fuels. For these countries, an increased share of electricity in the final energy mix has placed an upward pressure on $CO_2$ emissions. In contrast, for those countries in which electricity production is predominantly nuclear or hydro-based, the growing share of electricity actually reduced overall emissions. Many IEA countries experienced a downward trend in $CO_2$ emissions intensity due to a combination of cleaner fuels and the use of more efficient plants for the production of electricity and heat (Figure 2.8).

**Figure 2.8 ▶** *Trends in the $CO_2$ Intensity of Electricity and Heat Production*

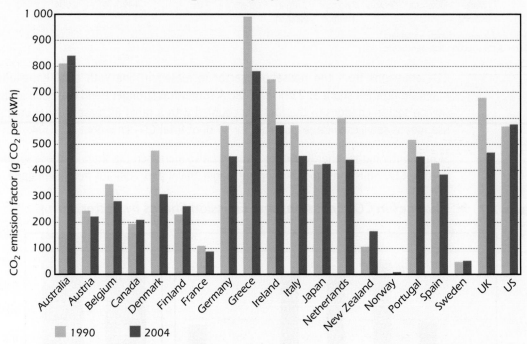

1990    2004

Note: Due to the difficulty of calculating separately the intensities of electricity and heat, the combined $CO_2$ intensity of electricity and heat production is shown for each country.
Source: IEA/OECD *$CO_2$ Emissions from Fuel Combustion.*

Norway has the lowest $CO_2$ intensity of electricity and heat production due to the dominance of hydro-electricity production, which is $CO_2$-free. Significant contributions from hydro and/or nuclear place France and Sweden well below the IEA average for $CO_2$ intensity. By contrast, the vast majority of electricity in Australia and Greece is produced from coal, making the carbon intensities of these countries among the highest in the IEA. Substantial declines in the carbon intensity can be seen for two countries in particular. In the United Kingdom, the decline was linked to a change in the fuel mix for electricity production from coal to gas. In Denmark, two factors were largely responsible: increased electricity production from wind and other renewables and efficient combined heat and power plants generating a larger share of both electricity and heat.

# Drivers of Energy Use

Energy underpins nearly every aspect of a modern economy. Thus, increases in economic activity in all sectors tend to put an upward pressure on energy consumption. Figure 2.9 shows the evolution of some of the key drivers of energy consumption across each sector over the period 1990 to 2004, on a per capita basis.

**Figure 2.9** ▶ *Changes in Key Sub-sectoral Activity Levels Relative to Population, 1990 - 2004*

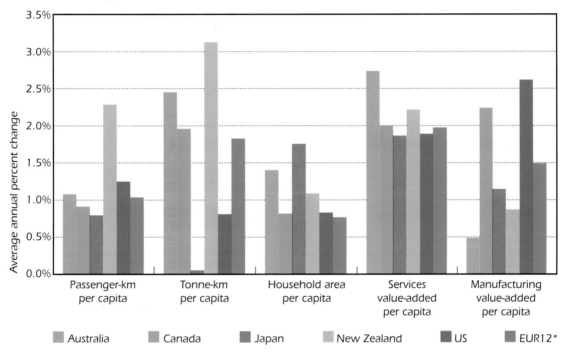

*EUR12 includes Austria, Denmark, Finland, France, Germany, Greece, Italy, the Netherlands, Norway, Spain, Sweden and the United Kingdom.

All drivers shown have increased between 1990 and 2004, with annual growth rates of between 1% and 2% for the IEA14 as a whole. The service sector showed particularly strong increases in value-added output per capita for all countries, which is one of the reasons why energy use in services is growing most rapidly. With the exception of Australia and Japan, household area per capita has grown at similar rates across many IEA countries. Other sectors demonstrate more significant differences among countries, with freight transport and manufacturing having the widest variation in activity increases. Freight haulage in Japan showed particularly low growth rates compared to other countries; consequently, energy use for freight transport in Japan was virtually unchanged between 1990 and 2004. In manufacturing, North America has experienced much higher growth rates per capita than in either Japan or Europe. However, the impact of this growth on energy use has been largely offset by significant intensity reductions.

Wide variations in some of the key determinants of energy use can be attributed to different growth rates among IEA countries since 1990, as well as a range of starting levels. Figure 2.10 shows the spread of several factors that are important for understanding differences in energy consumption per capita among the IEA14 countries. A level of 100% on the graph represents the weighted average value for a particular factor for the IEA14. The range, depicted by the bars, shows how the maximum and minimum values for a country compare to this average. Overall, there has been little change in the spread of most of these factors over time. However, it is interesting to note that the extent of differences across countries varies depending on the factor being considered. For most of the economic parameters – such as GDP per capita or value-added per capita for manufacturing and services – the differences among countries are less than a factor of two. This reflects the fact that all the IEA14 are modern, industrialised economies at similar stages of economic development.

The range in residential floor area per capita is somewhat wider, although it narrowed slightly between 1990 and 2004 to be just over a factor of two. The per capita size of dwellings is one of the key factors affecting energy use in households, together with climatic differences. These are represented on the chart by the variation in heating degree-days, which was just over a factor of three – from 1 500 in Japan to 4 900 in Finland.

**Figure 2.10 ▶ _Variation in Key Factors Affecting Energy Use, IEA14_**

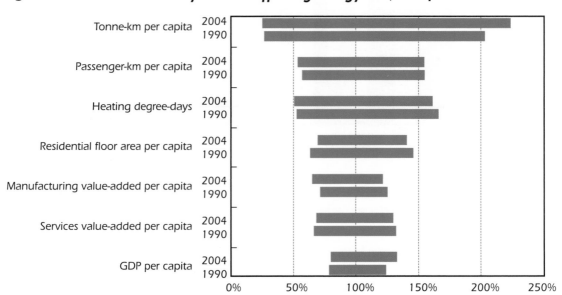

The widest differences among countries are seen for transport. The size of a country has a large impact on the extent of travel and freight haulage per capita. Variations in passenger travel per capita are about a factor of three; those for freight haulage per capita are even higher, at nearly nine.

The evolution of fossil fuel prices is a further important factor affecting energy consumption (Figure 2.11). These prices are strongly influenced by trends in spot oil prices (Figure 2.12). Most of the 1990s was characterised by decreasing real prices of crude oil. With the brief exception of the first Gulf crisis, real crude oil prices remained between USD 15 and USD 35 per barrel for 15 years (1988 – 2003).

Real prices for natural gas and, to some extent, coal have followed oil price developments, although with less significant fluctuations. Real energy prices decreased between 1990 and 1998; in many cases, they did not return to their 1990 levels until around 2004. Such price developments are unlikely to have provided much incentive for improvements in energy efficiency during the 1990s. Since 2003 prices have risen more steeply and these increases could be expected to have a significant effect on end-use energy efficiency and end-use fuel mix. However, detailed energy consumption data for years after 2004 were not available to confirm this at the time of writing.

**Figure 2.11 ▶** *Fossil Fuel Import Prices in Real Terms*

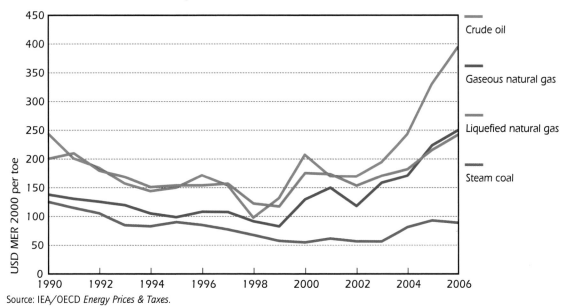

Source: IEA/OECD *Energy Prices & Taxes*.

**Figure 2.12 ▶** *Evolution of Crude Oil Prices*

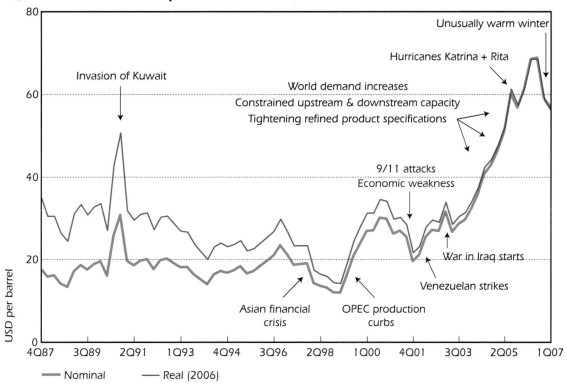

Source: IEA/OECD *Energy Prices & Taxes*.

# Changes in Aggregate Energy Intensity

The ratio of TFC per unit of GDP provides a measure of aggregate final energy intensity for a country, and is one of the most frequently used energy indicators. The calculation is typically made using GDP expressed either in purchasing power parity (PPP) or at market exchange rates (MER). Box 2.2 explains more about these two alternate measures of GDP and describes how they affect the calculation of energy intensity.

Both calculation methods result in a considerable spread in the final energy intensities of IEA countries (Figure 2.13). Indeed, the variation among countries is greater when GDP is calculated using MER rather than PPP. For example, using a PPP approach Canada consumed 2.5 times more energy per GDP in 2004 than Ireland. Calculated on a MER basis, the final energy intensity of Canada in 2004 was more than 3.5 times that of Japan. Several factors explain why these variations in energy consumption levels per unit of economic output are so different, even for countries that are at similar stages of development. Part of the difference reflects variations in energy efficiency. However, it would be misleading to rank energy-efficiency performance according to a country's energy consumption per GDP measured using either PPP or MER. The ratio is affected by many non-energy factors such as climate, geography, travel distance, home size and manufacturing structure.

Trends in final energy intensities show that the ratio of TFC to GDP (whether measured at PPP or MER) has declined in all IEA14 countries since 1990. However, the rate at which it has fallen varies. Norway had the strongest decline of the IEA14, with the ratio of TFC to GDP falling by 24% between 1990 and 2004, mostly due to changes in the structure of the economy.[4] At the other end of the scale, Italy had a decline of only 2%. For many countries, the TFC/GDP ratio fell most rapidly during the mid- to late 1990s, during a period of rapid economic growth.

$CO_2$ emissions per unit of GDP (whether calculated using PPP or MER) have also declined across all countries, reflecting both reductions in final energy use per GDP and changes in the final energy mix (Figure 2.14). Even more striking is the variation in levels of $CO_2$ emissions per GDP. When GDP is calculated using PPP, the emissions varied by more than a factor of three across the countries in 2004. This increased to a factor of more than four when using MER. The variation in emissions per GDP is greater than that for final energy consumption per GDP, with most of the differences being in the stationary sectors.

These findings indicate that the energy mix, especially for electricity generation, is an important determinant of $CO_2$ emissions (see Box 2.1). For example, Norway has a very high share of electricity in the final energy mix, but most of its electricity supply is hydro-based and has almost no $CO_2$ emissions. Thus, it is not surprising that emission levels per GDP from Norway's stationary sectors are low, despite a very energy-intensive manufacturing structure and a cold climate. Emissions from the stationary sectors are also low in other countries in which hydro and/or nuclear dominate electricity generation (*e.g.* France and Sweden).

---

4. Ireland (which was excluded from the IEA14 due to a lack of data for some sectors) experienced an even stronger decline in final energy intensity of 36%.

**Figure 2.13** ▶ *Final Energy Use per Unit of GDP*

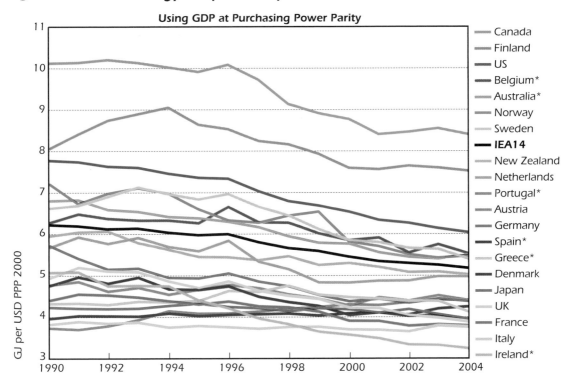

Using GDP at Purchasing Power Parity

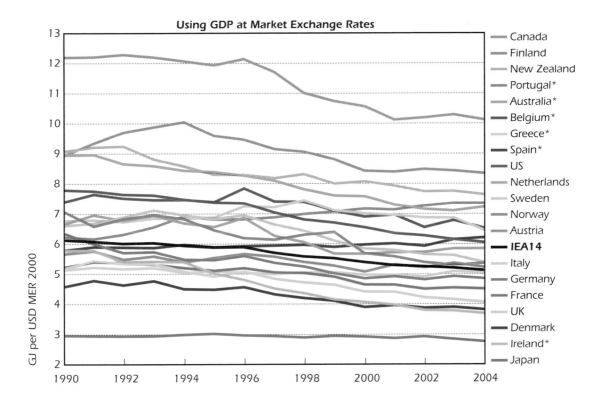

Using GDP at Market Exchange Rates

*Not included in IEA14.

**Figure 2.14 ▶** *CO₂ per Unit of GDP and Sector, 2004*

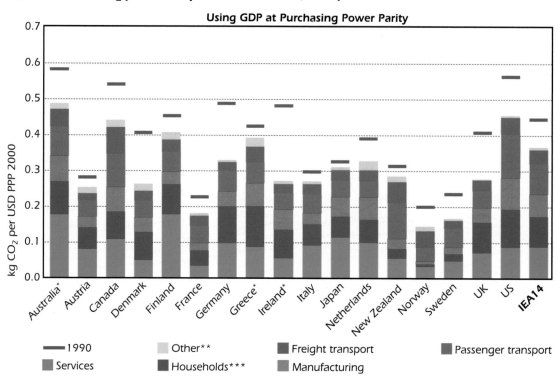

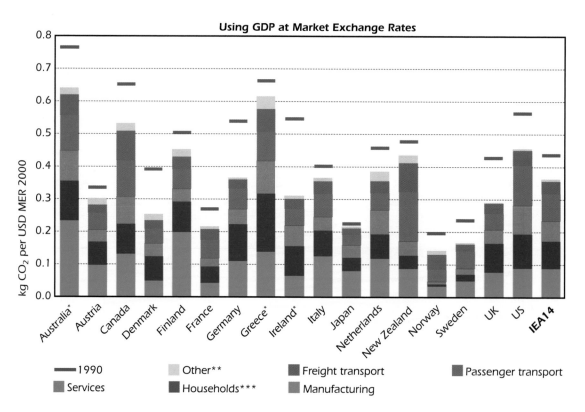

*Not included in IEA14.
**Other is construction and agriculture.
***Household CO₂ emissions are corrected for yearly climate variations, except in the case of Australia and Ireland.

In contrast, Australia, Canada and the United States have high emissions per unit of GDP because of their energy-intensive economies. In the case of Australia and the United States, high emissions also result from the fact that a significant proportion of electricity production is coal based. A large proportion of coal-based electricity also gives Greece a higher than average $CO_2$ intensity, particularly when GDP is measured in MER. Thanks to the high share of low-carbon electricity in Norwegian space heating, the two building sectors (households and services) have almost no $CO_2$ emissions. Like Norway, Finland has a very cold climate. However, Finland relies much more heavily on fossil fuels for both electricity production and space heating; as a result, Finland has much higher $CO_2$ emission levels from buildings in the household and service sectors. Denmark, Germany and the United Kingdom also have relatively high building sector emissions. For these countries, the large share of coal in the electricity mix is a more significant factor than a cold climate.

The passenger and freight transport sectors are almost entirely based on oil products. The more modest variations in $CO_2$ emissions per GDP levels in these sectors reflect differences in transport distances per unit of GDP, energy intensities and – for some countries – differences in the mix of transport modes. Australia, Canada, New Zealand and the United States have the highest $CO_2$ emissions per GDP from passenger travel. In most cases, this can be attributed to relatively high energy intensities for cars combined with long driving distances – the latter being at least partly due to geography.

---

**Box 2.2**

## Alternative Approaches for Calculating Energy and $CO_2$ Intensities

Energy and $CO_2$ intensities are calculated by dividing energy use and $CO_2$ emissions, respectively, by an economic measure of output such as GDP or value-added. A common unit to measure output (*e.g.* in USD) must be used to compare results across countries: the two main approaches for this conversion are market exchange rates (MER) and purchasing power parity (PPP).

A MER approach simply uses actual exchange rates to convert GDP or value-added in national currencies to a common currency, such as USD. In contrast, the PPP approach defines a "basket of goods" (or services) and then equalises the purchasing power of various currencies to "buy" these goods in their home countries. These special exchange rates are then used to convert GDP or value-added to USD. The two approaches produce different results, as is evident in Figure 2.15, which compares final energy use per unit of GDP for 2004. This variation can affect how countries compare with one another.

In reality, MER can fluctuate widely. Thus, it is often argued that PPP exchange rates better reflect the long-run equilibrium value between currencies. Furthermore, as PPP measures of output are not affected by different price levels, they should be more closely linked to the level of physical activity in an economy.

However, it should be recognised that the use of PPP has its own problems. Calculating PPP is complicated by the fact that countries do not simply differ in a uniform price level. The difference in energy prices between countries may be greater than the difference in housing prices and, at the same time, less than the difference in raw material prices. In addition, people

**Box 2.2** continued

and businesses in different countries typically consume different baskets of goods. To compare effectively the cost of baskets of goods and services, it is necessary to create a price index. This is a difficult task because purchasing patterns – and even the goods available to purchase – differ from country to country. In addition, adjustments must be made to reflect differences in the quality of goods and services.

**Figure 2.15** ▶ *Alternative Approaches for Calculating Final Energy Intensity, 2004*

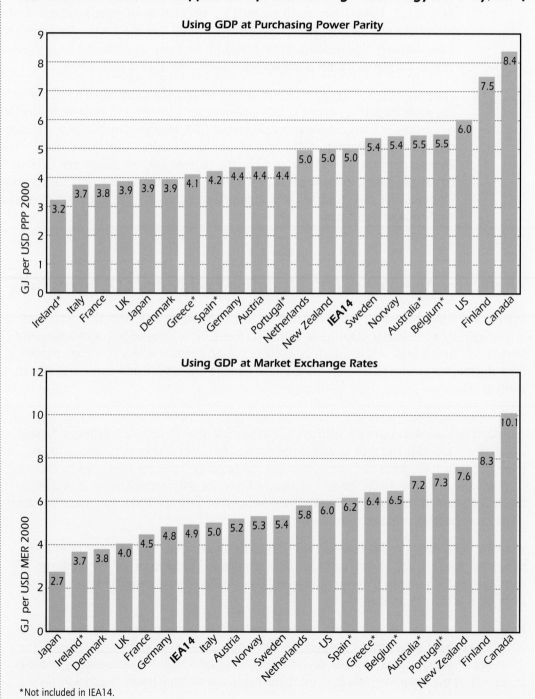

*Not included in IEA14.

**Box 2.2** continued

Despite its complications and limitations, PPP is used for many international comparisons. In keeping with this practice, most of the energy and $CO_2$ intensities in this publication are calculated using PPP to convert national currencies into USD. Some key indicators have been calculated using both PPP and MER measures of GDP or value-added; where this is the case, it is explicitly noted. Readers are reminded that the two calculations often result in different absolute values for the indicators and may also change the position of countries relative to each other. However, the trends over time for individual countries remain the same.

**2**

# Impact of Changes in Energy Services, Structure and Intensity

The previous section provided evidence of the large variations among countries in how much energy use per unit of GDP has fallen over time. To understand the extent to which this variation reflects differences in energy-efficiency developments, it is necessary to separate the impact of changes in sub-sectoral energy intensities (which are used as a proxy for energy efficiency) from the effects of changes in economic structure and other factors that influence the demand for energy (such as those depicted in Figure 2.10).

Increased demand for energy services – reflecting increased ownership levels of electric household appliances, bigger houses, more personal travel by car, etc. – drives both energy use and energy use per GDP. Therefore, it is useful to examine how the energy per GDP ratio is affected by changes in the ratio of energy services to GDP and in end-use intensities (such as the energy used to heat a square metre of floor space or to ship a tonne of freight one kilometre). This is done using an index decomposition method, which is briefly described in Box 2.3.

Changes in energy consumption per GDP can be attributed to changes in the ratio of energy services to GDP and to changes in sub-sector energy intensities for more than 20 end-uses (Figure 2.16). The intensity effect for the whole economy is calculated as the aggregate impact of the sectoral intensity effects.

The results of aggregate impact calculations show that the energy intensity effect and the decoupling of energy service demand and GDP have both contributed to reduced energy consumption per unit of GDP. However, declining end-use intensities (the energy intensity effect) have been the most important factor. In the early 1990s, reductions in energy services per GDP made almost no contribution to the falling energy to GDP ratio; the economy of the IEA was growing slowly, as were demands for energy services. The contribution from a reduction in energy services per GDP has grown since then, but by 2004, some 66% of the total decline in energy per GDP could be attributed to reductions from the energy intensity effect. This equates to an annual average fall in end-use intensities across all sectors of 0.9% per year. Still, this rate of reduction is substantially less than in the 1973 – 90 period, when the fall in end-use intensities averaged 2% per year (IEA, 2004).

**Box 2.3**

## *Decomposing Changes in Energy Use*

The IEA methodology for analysing energy end-use trends distinguishes between three main components affecting energy use: activity levels, structure (the mix of activities within a sector) and energy intensities (energy use per unit of sub-sectoral activity). Depending on the sector, activity is measured as value-added, passenger-kilometres, tonne-kilometres or population. Structure further divides activity into industry sub-sectors, transportation modes, or measures of residential end-use activity. Using an appropriate measure of end-use activity, energy intensities are then calculated for each of these sub-sectors, modes or end-use activities.

The energy intensity effect, which is used as a proxy for changes in energy efficiency, separates out how changing energy intensities influence energy consumption for a particular sector. This is done by calculating the relative impact on energy use that would have occurred between a base year (usually 1990 in this publication) and a future year (usually 2004) if the aggregate activity levels and structure for a sector remained fixed at base year values while energy intensity followed its actual development. A similar approach is used to calculate the activity and structure effects, which together represent the energy service effect. (Annex A provides further details of the decomposition approach.)

The separation of impacts on energy use from changes in activity, structure and intensity is critical for policy analysis. Most energy-related policies target energy intensities and efficiencies, often by promoting new technologies. Accurately tracking changes in intensities helps measure the effects of these new technologies.

**Figure 2.16 ▶** *TFC per GDP and the Contribution of Changes in Energy Services per GDP and Energy Intensities, IEA14*

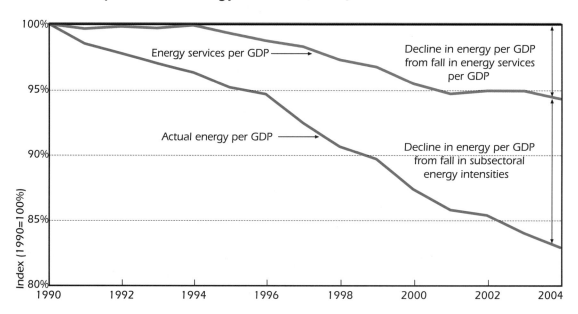

The relative contribution of changes in energy services per GDP and the intensity effect to the overall trend varies among countries (Figure 2.17). With the exception of Japan, all countries show that the intensity effect contributed to reducing the ratio of energy use to GDP: for most countries, it was the dominant factor. This is particularly true in the case of Canada, Finland, Germany, New Zealand, Sweden and the United States. In contrast, for Norway and the United Kingdom, changes in energy services per unit of GDP were most important. For those countries having declining sub-sectoral intensities, it is possible to identify three groups according to the magnitude of the intensity effect:

▷ Less than 0.5% per year: Austria, Finland, Italy, the Netherlands, and the United Kingdom.

▷ 0.5% – 1%: Denmark, France, Norway and Sweden.

▷ Greater than 1%: Canada, Germany, New Zealand and the United States.

**Figure 2.17 ▶** *Changes in TFC/GDP Decomposed into Changes in Energy Services/GDP and Intensity Effect, 1990 – 2004*

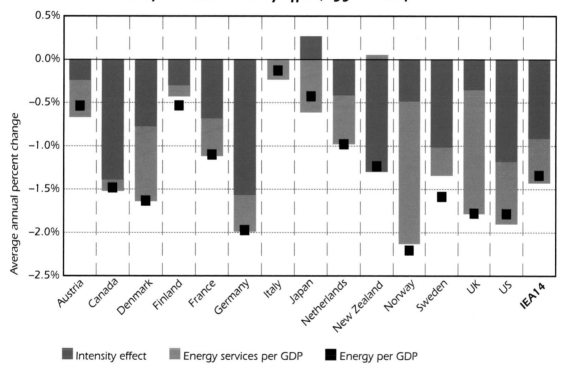

It can be informative to compare these results with the information on final energy intensity per GDP shown in Figure 2.13. Such a comparison shows that those countries with a high level of energy use per GDP in 1990 tend to have had the largest reductions in intensity. In contrast, those countries that initially had lower energy use per GDP have generally seen smaller declines in intensity.

Although the intensity effect for the whole economy declined by an average of 0.9% per year, the rate of reduction was not uniform across end-use sectors. The manufacturing sector showed the largest reductions in intensities (corrected for structure), decreasing by 1.4% per year between 1990 and 2004 (Figure 2.18). In the later years, the rate of reduction appears to be increasing, but is still substantially lower than the 2.8% reduction per year seen in the 1973 – 90 period (see Annex C for a discussion of long-term trends). Building energy use showed lower reductions, with average intensity falls of 0.8% per year in households and 1.2% in the service sector. There is no significant discernable trend in the rate of reductions – except that, again, these rates are much lower than seen prior to 1990 when the average annual reductions were 2.0% for households and 2.3% for services. The passenger and freight transport sectors experienced similar declining trends between 1990 and 2004, but had the lowest falls in intensities, at 0.5% and 0.6% per year, respectively (calculated holding the modal mix constant). This compares with rates between 1973 and 1990 of 1.0% per year for passenger transport and 0.7% for freight transport.[5]

**Figure 2.18** ▶ *Sector Intensities and Total Economy Effect, IEA14*

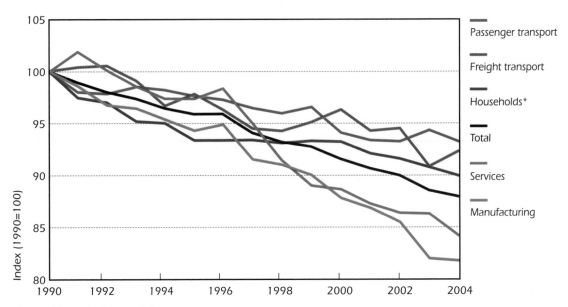

*Corrected for yearly climate variations.

Examining the three effects discussed in Box 2.3 – activity, structure and intensity – makes it possible to analyse in more detail how the factors affecting total final energy consumption have evolved over time (Figure 2.19). The combined impact of the activity and structural effects is equal to the energy service effect previously discussed. In the early 1990s, GDP growth was relatively low at 2% per year. As a result of the combined impacts of activity and structure, demand for energy services was growing at almost the same rate (even though the role of the structure effect was very small). This increase in the demand for energy services was partially offset by a decline in the intensity effect. As a result, final energy use increased by an average of 1% per year.

5. The figures for the declines in energy intensity between 1990 and 2004 presented here may differ from those in the individual sector chapters due to differences in the IEA countries included in the calculations.

In the mid- to late 1990s, economic growth accelerated. The demand for energy services also increased more rapidly. There was some increase in the rate of energy intensity reduction during this period, but it was not sufficient to prevent the rate of final energy demand growth rising to an average of 1.4% per year. After 2000, economic growth and the demand for energy services again slowed; the structure effect became negative. This slowing of underlying service demand, coupled with a further increase in the rate of energy intensity reduction, was sufficient to keep the growth in final energy use to below 0.5% per year.

**2**

**Figure 2.19** ▶ *Factors Affecting Total Final Energy Consumption, IEA14*

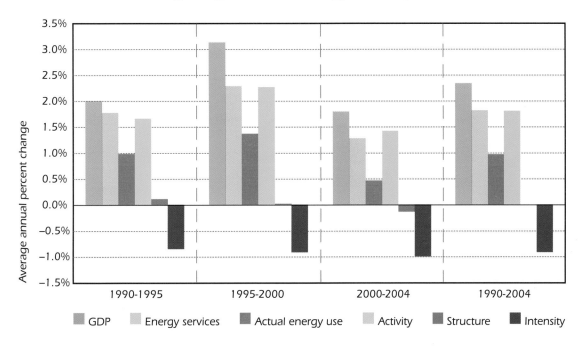

In summary, there are a number of important observations in relation to the factors affecting the change in energy use and energy use per GDP since 1990. First, the demand for energy services has consistently increased at a slower rate (1.8% per year) than GDP (2.3% per year), although the gap narrows during periods of recession. Second, the decline in energy use per GDP can mostly be attributed to falling sub-sectoral intensities, which are used as a proxy for improvements in energy efficiency. The third observation is that the rate of decline in sub-sectoral intensities (the energy intensity effect) appears to be increasing slightly. However, as noted earlier, at an average rate of less than 1% per year, it is still well below the rates seen during the 1973 – 90 period. Fourth, final energy use continued to grow in IEA countries, at an average rate of almost 1% per year between 1990 and 2004.

Figure 2.20 shows how changes in the energy mix and the carbon intensity of energy commodities have affected the development of $CO_2$ emissions. $CO_2$ emissions have increased between 1990 and 2004 but, on average, at a very slightly slower rate than the increase in final energy use. This is the result of two opposing trends. Changes in the end-use energy mix – particularly the increased share of electricity –

have tended to increase emissions. However, this effect has been more than offset by reductions in the carbon intensity of individual energy commodities, with electricity again playing a key role (see Box 2.1).

The way these factors have affected overall $CO_2$ emissions has evolved over time. Between 1990 and 1995, changes in both end-use fuel mix and carbon intensity tended to slow the rate of increases in $CO_2$ emissions, relative to energy consumption. During the next five-year period (1995 – 2000), $CO_2$ emissions actually increased more quickly than final energy consumption because the end-use fuel mix effect reversed. After 2000, the end-use energy mix effect continued to exert an upwards pressure on $CO_2$ emissions. However, this was just offset by a decline in the carbon intensity of electricity production. Consequently, $CO_2$ emissions grew very slightly less quickly than final energy use.

**Figure 2.20** ▶ *Decomposition of Changes in $CO_2$ Emissions, IEA14*

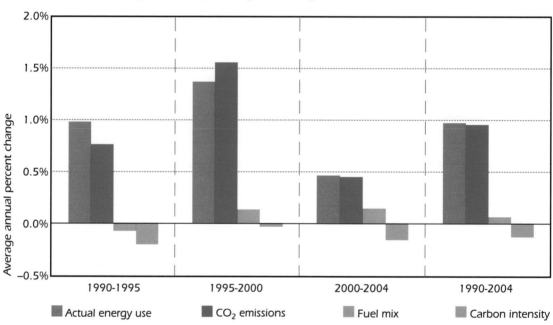

## Energy and $CO_2$ Savings

The decline in energy intensities (*i.e.* improvements in energy efficiency) in the various end-use sectors led to energy and $CO_2$ savings across the whole economy (Figure 2.21). Energy savings in 2004 were almost 16 EJ (370 Mtoe) or 14% of total final energy use in that year, which represents an estimated USD 170 billion of energy cost savings. The impact of energy intensity improvements on $CO_2$ emissions was also significant, with a saving of 1.2 Gt $CO_2$ (14% of actual $CO_2$ emissions in 2004). These savings are approximately equal to the annual final energy consumption and $CO_2$ emissions of Japan.

Figure 2.22 shows the total energy savings since 1990, broken down by sector. In 2004, the largest share was from manufacturing (43%), followed by households and services (18% and 19%, respectively), passenger transport (14%) and freight transport (6%). Initially, there were strong energy savings in households as a result of significant improvements in space heating intensity, which did not continue. In contrast, the savings from services made an impact only in the late 1990s, during a period of high economic growth in this sector. The rates of savings in the other sectors were more constant over time.

**Figure 2.21** ▶ *Energy and CO2 Emissions Savings from Improvements in Energy Intensity, IEA14*

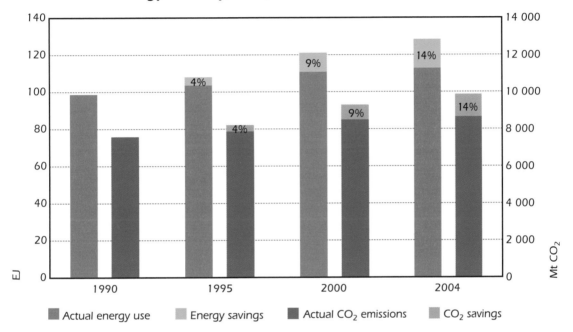

**Figure 2.22** ▶ *Contribution to Energy Savings from the End-use Sectors, IEA14*

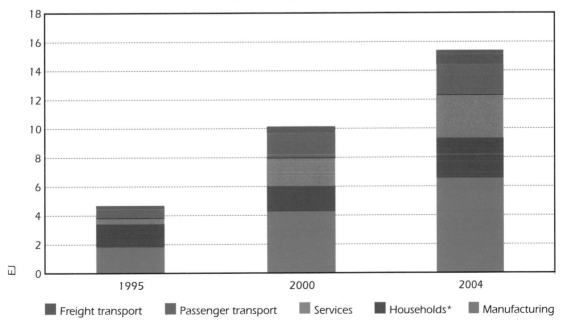

*Corrected for yearly climate variations.

# Summary

An examination of economy-wide trends in energy use and $CO_2$ emissions in 14 IEA countries over the period 1990 to 2004 reveals the following key findings:

▷ Total final energy use increased by 14% between 1990 and 2004. Oil is still the most important final energy commodity, with a share of 47% in 2004. The transport sector now accounts for over 75% of total final oil use. Electricity overtook natural gas as the second most important final energy commodity; its use increased rapidly in all stationary sectors, with the strongest growth in households and services. Use of combustible renewables (with exception of biofuels) has not significantly increased from 1990 levels; therefore, its share in end-use consumption remained constant.

▷ $CO_2$ emissions in the IEA14 have increased in line with energy use, also showing a 14% rise between 1990 and 2004. Emissions are increasing in all major end-use sectors except manufacturing. Despite this, manufacturing has the highest share (24%) of total IEA14 emissions in 2004.

▷ GDP across the IEA14 increased, on average, by 2.3% annually. Demand for energy services in the economies of IEA countries is increasing more slowly than GDP, at 1.8% per year. There are wide variations among countries in the levels and rates of change of many of the key factors affecting final energy consumption.

▷ Real energy prices decreased between 1990 and 1998; in many cases, they did not return to their 1990 levels until around 2004. Such price developments are unlikely to have provided much incentive for improvements in energy efficiency.

▷ Improvements in energy efficiency led to a reduction in end-use energy intensities of 0.9% per year, around half the rate seen in previous decades. The manufacturing sector showed the largest reduction in energy intensity, decreasing by 1.4% per year. Building energy use showed smaller reductions of 0.8% per year in households and 1.2% per year in the service sector. Transport showed the lowest declines in intensity with passenger and freight transport intensities falling by 0.5% and 0.6% per year, respectively.

▷ Reductions in end-use intensities were the main reason behind a decline in energy use per GDP in the IEA14. In general, countries that had high levels of final energy use per GDP in 1990 have since experienced the most significant reductions in intensity.

▷ Without the energy savings resulting from reductions in energy intensity, total final energy consumption in the IEA14 would have been 14% higher in 2004. This represents an annual energy saving in 2004 of 16 EJ, worth USD 170 billion, and 1.2 Gt of avoided $CO_2$ emissions.

▷ Projections published by the IEA in *Energy Technology Perspectives 2006* (IEA, 2006a) show that the recent rate of improvement in energy efficiency would need to at least double for a realistic chance of a more sustainable energy future. This is indeed possible; the Alternative Policy Scenario of the *World Energy Outlook 2006* (IEA, 2006b) shows that implementation of the energy-efficiency policies currently under consideration by OECD countries could reduce final energy consumption by more than 7% in 2030, with net financial savings.

# MANUFACTURING

## Scope

The manufacturing sector of industry produces finished goods and products for use by other businesses, for sale to domestic consumers or for export. Total manufacturing is divided into the following key industries:

▷ primary metals

▷ chemicals

▷ non-metallic minerals

▷ paper, pulp and printing

▷ food, beverages and tobacco

▷ metal products and equipment

▷ other manufacturing.

Fuel-processing industries and fuels used as feedstocks are not included in the analysis.

## *Highlights*

Between 1990 and 2004, the overall energy efficiency of manufacturing industry in a group of 19 IEA countries improved by 1.3% per year. Without the energy savings resulting from these improvements, manufacturing energy consumption in the IEA19 would have been 21% higher in 2004 (Figure 3.1). This represents an annual energy saving of 6.7 EJ in 2004, which is equivalent to 490 Mt of avoided $CO_2$ emissions.

The effect of these savings is significant: despite a 31% increase in output, final energy use in the manufacturing sector increased by only 3% between 1990 and 2004. However, the rate of improvement in energy efficiency during this period was much lower than in previous decades. Between 1973 and 1990, the average annual rate of improvement was 2.8%.

**Figure 3.1** ▶ *Manufacturing Energy Savings from Improvements in Energy Efficiency, IEA19*

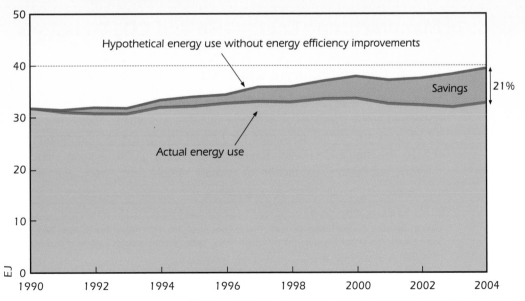

# Overview of Trends in the Manufacturing Sector

Within a group of 19 IEA countries for which comprehensive data are available[1], recent trends (1990 – 2004) in manufacturing show a marked decoupling of economic activity and energy use (Figure 3.2). Over this time, output from manufacturing (as measured by value-added) increased by 31%. In contrast, total final energy use increased by only 3%. This decoupling resulted in a 22% decrease in final energy intensity. $CO_2$ emissions, including those from the production of electricity and heat used in the manufacturing sector, rose by only 1%, reflecting a decrease in the carbon intensity of the energy mix.

**Figure 3.2** ▶ *Overview of Key Trends in Manufacturing, IEA19*

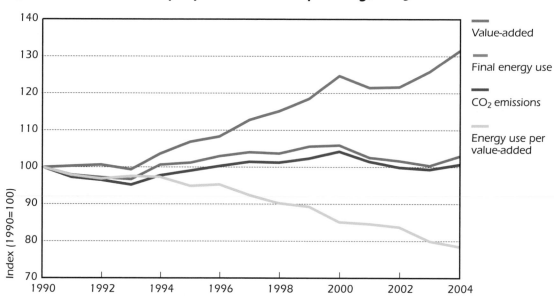

# Trends in Manufacturing Energy Use and $CO_2$ Emissions

### Energy Consumption by Sub-sector

The majority of energy use in manufacturing is used to produce raw materials such as paper and pulp[2], chemicals, non-metallic minerals and primary metals (Figure 3.3). In 1990, these sub-sectors accounted for 72% of energy use in manufacturing in the IEA19. This figure decreased slightly, to 69%, by 2004. Among these energy-intensive sub-sectors, production of primary metals (iron, steel, aluminium, etc.)

---

1. The 19 IEA countries included in the analysis of the manufacturing sector are Australia, Austria, Belgium, Canada, Denmark, Finland, France, Germany, Greece, Italy, Japan, the Netherlands, New Zealand, Norway, Portugal, Spain, Sweden, the United Kingdom and the United States. These countries account for 91% of total manufacturing energy use in all IEA countries.

2. In this publication, energy and value-added data for the energy-intensive paper and pulp sub-sector also includes the much less energy-intensive activity of printing due to the difficulty of separating out printing using IEA and OECD statistics.

continues to consume the largest share of total energy in manufacturing (24%), followed by chemicals (19%), and paper and pulp (17%). Despite the small decrease in its share of manufacturing energy use, the absolute level of energy consumption for the production of raw materials in the IEA19 remained largely constant from 1990 to 2004.

**Figure 3.3** ▶ *Energy Use by Manufacturing Sub-sector, IEA19*

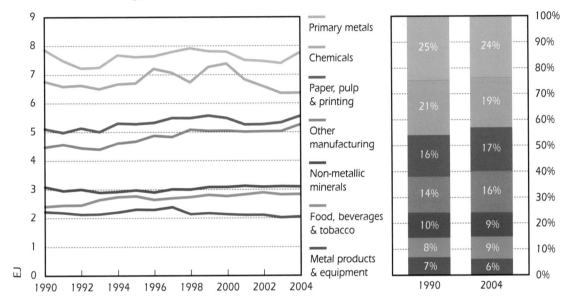

## Energy Consumption by Fuel

Total manufacturing energy use in the IEA19 increased by 3% between 1990 and 2004. The small increase in overall energy use in manufacturing masks some important changes in the final energy mix – particularly, that consumption of oil and coal declined while natural gas and electricity use increased (Figure 3.4). These developments continue longer term trends that date back to the early 1970s. In 2004, the share of oil was 13%, down from 15% in 1990; coal had declined to 19% from the 1990 share of 25%. In contrast, natural gas continued to be the most important energy commodity in the manufacturing sector, increasing its share of final energy use from 27% in 1990 to 29% in 2004. Electricity use increased more significantly; its share rose from 24% to 27%. The shares of both renewables and district heat also increased, although more modestly.

Changes in the manufacturing fuel mix, including the trend away from coal and oil towards natural gas and electricity, can be largely attributed to three factors: changes in relative fuel prices, shifts in industry structure and processes, and the implementation of environmental legislation that favours the use of cleaner fuels. Despite having no emissions at the point of use, the increased share of electricity in the manufacturing fuel mix has tended to increase $CO_2$ emissions because many IEA countries rely on fossil fuels for electricity generation (see Box 2.1 in Chapter 2).

**Figure 3.4** ▶ *Manufacturing Energy Use by Fuel, IEA19*

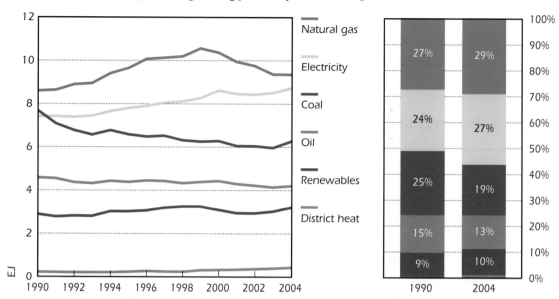

## CO₂ Emissions

A combination of two factors determines each sub-sector's contribution to total manufacturing $CO_2$ emissions: the amount of energy used and the end-use fuel mix (non-energy, process-related $CO_2$ emissions are not included in this analysis). Primary metals and chemicals are the largest contributors to $CO_2$ emissions in manufacturing (Figure 3.5). In contrast, although paper and pulp accounts for a large share of energy use, this sub-sector produces a lower share of $CO_2$ emissions. This is because biomass, which has no net $CO_2$ emissions, makes up a substantial amount of the paper and pulp fuel mix.

**Figure 3.5** ▶ *CO₂ Emissions by Manufacturing Sub-sector, IEA19*

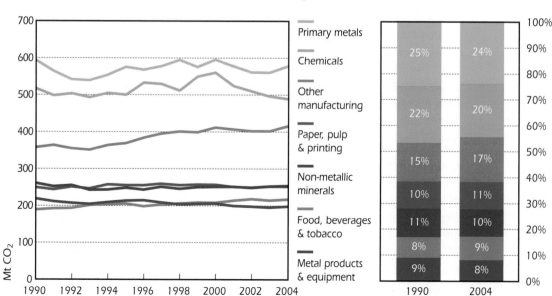

| Box 3.1 |

# Energy and $CO_2$ Indicators for Industry:
## Using Economic or Physical Ratios

Energy and $CO_2$ indicators for manufacturing industry typically comprise a measure of energy use or $CO_2$ emissions divided by a measure for activity. Activity is usually defined as the production of a given sector or product, and can be measured in either economic terms (*e.g.* value-added) or in physical units (*e.g.* weight or number of products). Generally, indicators calculated in monetary units are used at a sectoral or economy-wide level. Indicators based on physical units are better suited to detailed sub-sectoral analyses.

The indicators for the manufacturing sector presented in this publication are economic ratios: they analyse energy use or $CO_2$ emissions per unit of value-added output. The advantages of this approach are two-fold.

First, the measure of activity (monetary value) is similar even though different products are produced. This makes it possible to compare indicators across different sub-sectors of manufacturing.

Second, on a practical level, value-added data for manufacturing industry – including data disaggregated into different sub-sectors – are available from the OECD for most IEA countries. This facilitates consistent cross-country comparisons. However, economic-based indicators suffer from one significant drawback: they are influenced by a range of pricing effects that are unrelated to changes in the level of underlying physical production.

An alternative approach is to calculate energy use and $CO_2$ emissions indicators based on physical ratios, *e.g.* using a measure of activity based on tonnes of production. These types of indicators are often called the "specific" or "unit energy" consumption; their advantage is that they are not affected by changes in prices. Thus, at a disaggregated level, physical indicators can give a better measure of the technical efficiency of a particular production process. However, because the denominator is measured as a physical unit, it is not possible to compare indicators defined in differing units without making conversions.

Data availability is also an issue for this approach. Limited data are available for some industries such as cement, iron and steel, aluminium, and paper and pulp; however, physical output data are not available for most industries. These factors make it difficult to use physical indicators to provide an aggregate picture of energy use and efficiency for the whole of manufacturing industry.

As both economic and physically based indicators have advantages and disadvantages, ideally, a combination of approaches should be used to examine energy use and $CO_2$ emissions in the manufacturing sector. The IEA recently published a major new analysis of industrial energy efficiency for a number of key sectors based on physical indicators, entitled *Tracking Industrial Energy Efficiency and $CO_2$ Emissions* (IEA, 2007). Future work will aim to combine this approach with the economic-based analysis presented in this chapter.

# Drivers of Energy Use in Manufacturing

Energy is a key input in the production of many manufactured goods. All other things being equal, an increase in manufacturing production (output) will generally lead to an increase in energy consumption.

In this study, manufacturing output is measured in economic rather than physical units. This makes it easier to add together the outputs from various industries (Box 3.1 examines in more detail the advantages and disadvantages of both economic and physical energy indicators). The measure of output chosen is "value-added", which reflects the difference between the value of the goods produced and the value of inputs.[3] Some industries, such as those producing raw materials, require more energy to produce one unit of value-added than others – *i.e.* they are more energy intensive. Thus, overall energy consumption derives from the total level of value-added output from the manufacturing sector and the relative contributions from the various industries.

For the IEA19, manufacturing output (value-added) increased by 31% between 1990 and 2004. However, this growth was not constant. The IEA19 experienced a period of recession in the early 1990s; as a result, manufacturing output changed very little between 1990 and 1993. A second recession occurred at the start of the current decade. In this case, output fell in real terms between 2000 and 2001, and was almost unchanged the following year (Figure 3.6). During the periods in which manufacturing expanded, its growth was in line with the rate of the overall economy. Overall, despite a significant increase of output in absolute terms, the share of manufacturing in total economic output fell slightly from 17.5% in 1990 to 16.5% in 2004.

The composition of manufacturing in the IEA19 changed gradually through the 1990s and into the early part of the current decade. In 1990, some 31% of total output derived from primary metals, chemicals, paper and pulp, and non-metallic minerals. By 2004, this share had declined to 29%. In contrast, there was a rapid increase in the share of several less energy-intensive sub-sectors, such as metal products and equipment.

IEA countries show significant differences in the structure of manufacturing industry (Figure 3.7). In several countries (including Australia, Canada, Belgium, Finland, the Netherlands, Norway and Sweden), more than one-third of total output comes from the production of raw materials. By contrast, in Japan and New Zealand these sub-sectors account for less than one-quarter of total manufacturing output. Most countries experienced a fall in the share of raw materials manufacturing. Particularly large reductions were evident in the Scandinavian countries of Finland, Norway and Sweden, as other parts of the manufacturing economy expanded rapidly. The share of raw materials in total output increased substantially in Australia and Belgium.

---

3. In this chapter, valued-added output for manufacturing was first calculated in national currencies at 2000 constant prices and then converted to USD using the relevant purchasing power parities for the year 2000 (see Box 2.2 in Chapter 2). Selected charts have also been re-calculated using value-added output converted to USD at 2000 market exchange rates; these are shown in Annex B.

**Figure 3.6** ▶ *Manufacturing Output, Shares of Manufacturing in Total GDP and of Raw Materials in Manufacturing Output, IEA19*

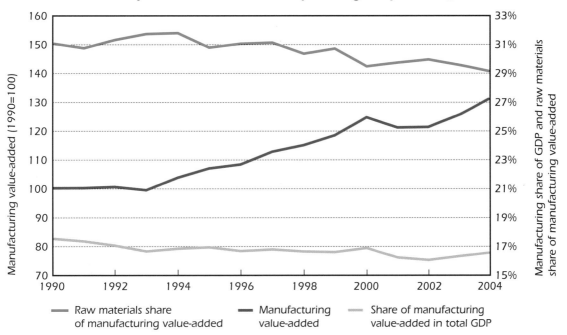

**Figure 3.7** ▶ *Composition of Manufacturing Value-added, IEA19*

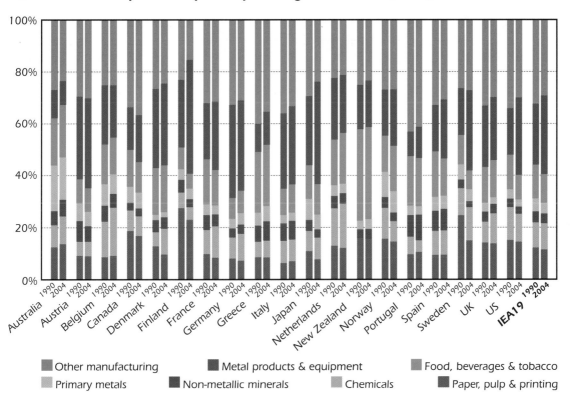

**3**

There is a significant spread in the level of manufacturing energy expenditure relative to value-added across IEA countries (Figure 3.8). This can be attributed to differences in the structure of manufacturing industry, together with variations in fuel prices. For some countries, a large share of energy-intensive raw material production leads to relatively high energy expenditure for manufacturing, even though fuels and electricity are not expensive compared to the IEA average. This is the case in Australia, Finland and Norway. In fact, access to cheap energy is often a stimulant for the production of energy-intensive materials. For example, in Australia and Norway, where electricity is relatively inexpensive, the production of aluminium – a very electricity intensive process – constitutes an important share of the production of primary metals and thus drives up the average intensity for this sub-sector.

**Figure 3.8** ▶ *Manufacturing Energy Expenditures Relative to Value-added, IEA19*

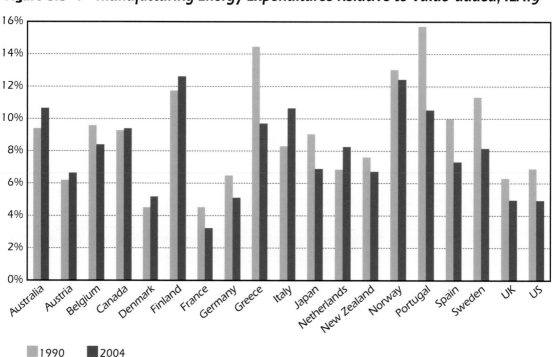

■ 1990   ■ 2004

In contrast, relatively low energy prices, combined with a smaller share of energy-intensive industries, leads to significantly lower manufacturing energy expenditures when compared to value-added. This reflects the situation in France, the United Kingdom and the United States.

Unlike many other end-use sectors, taxes generally comprise only a small proportion of energy prices in manufacturing (Figure 3.9). This is largely due to concerns that the unilateral imposition of higher taxes could disadvantage the economic competitiveness of a country's manufacturing industry within the global market place. Consequently, the levels of energy prices in this sector do not vary as significantly across countries as they do for households and transport.

**Figure 3.9** ▶ *Energy Prices in Manufacturing, 2004*

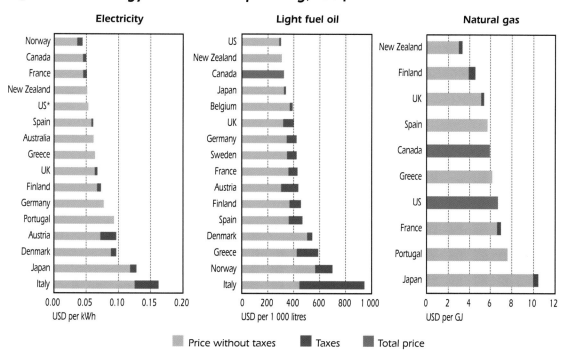

*Information on the tax component is not available.

Source: IEA/OECD *Energy Prices & Taxes.*

# Changes in Aggregate Energy Intensity

A measure of aggregate manufacturing energy intensity can be obtained by dividing total manufacturing energy use by total manufacturing value-added (Figure 3.10). For the IEA19, this intensity fell by 22% between 1990 and 2004, at an average rate of 1.7% per year. All IEA19 countries, except Spain, experienced a decline in this measure of energy intensity, with reductions of up to 40%. In the case of Spain, manufacturing energy use increased more quickly than value-added since the mid-1990s because of strong growth in a number of energy-intensive sub-sectors.

Large differences are evident in the level of aggregate manufacturing energy intensity among IEA countries. In general, three groups of countries can be defined.[4]

▷   High energy intensity countries: Australia, Canada, Finland and Norway.

▷   Medium energy intensity countries: Belgium, Greece, New Zealand, the Netherlands, Portugal, Spain, Sweden, and the United States.

▷   Low energy intensity countries: Austria, Denmark, France, Germany, Japan, Italy, and the United Kingdom.

The variations in energy intensity can be explained, at least to some extent, by differences in the structure of the sector – *i.e.* the mix of manufacturing products (see Box 3.2). These structural differences affect both the absolute levels of aggregate

---

4. Figure B.1 in Annex B presents manufacturing energy intensity calculated using market exchange rates. While there are some differences in the order of individual countries, the grouping of countries according to levels of energy intensity remains broadly unchanged.

intensities and the rates of decline. However, as Figure 3.11 illustrates, energy and value-added data at the sub-sector level are needed to quantify the importance of structural changes.

**Figure 3.10 ▶ *Energy Use per Unit of Manufacturing Value-added***

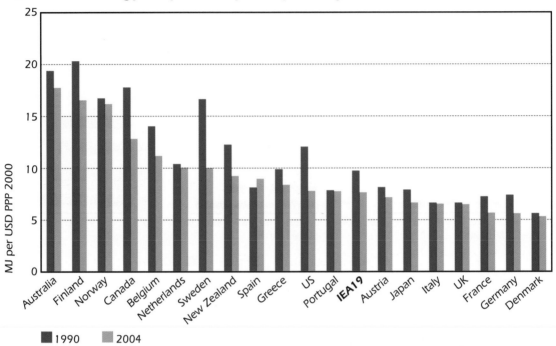

**Figure 3.11 ▶ *Sub-sector Energy Intensities, Value-added and Energy Shares, IEA19***

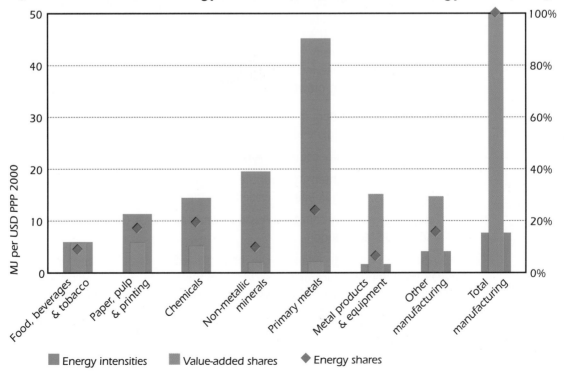

Box 3.2

# Impact of Structure on Manufacturing Energy Intensities

Large differences are evident in the level of aggregate manufacturing energy intensity among IEA countries. In addition, various sub-sectors of manufacturing have very different energy intensities. This raises an interesting question: *To what extent can differences in the energy intensity of manufacturing industry among countries be explained by differences in their industrial structure?*

To examine this question, one must remove the effects of structural differences by calculating what the energy intensity of each country would be if they all had a common structure (in this case, the average of the IEA19 countries, based on a breakdown of manufacturing into seven sub-sectors). Figure 3.12 shows the results of such a calculation for 2004, and compares these intensities at common structure with actual energy intensities in that year.

**Figure 3.12 ▶ *Manufacturing Energy Intensity at Actual and Common Structures***

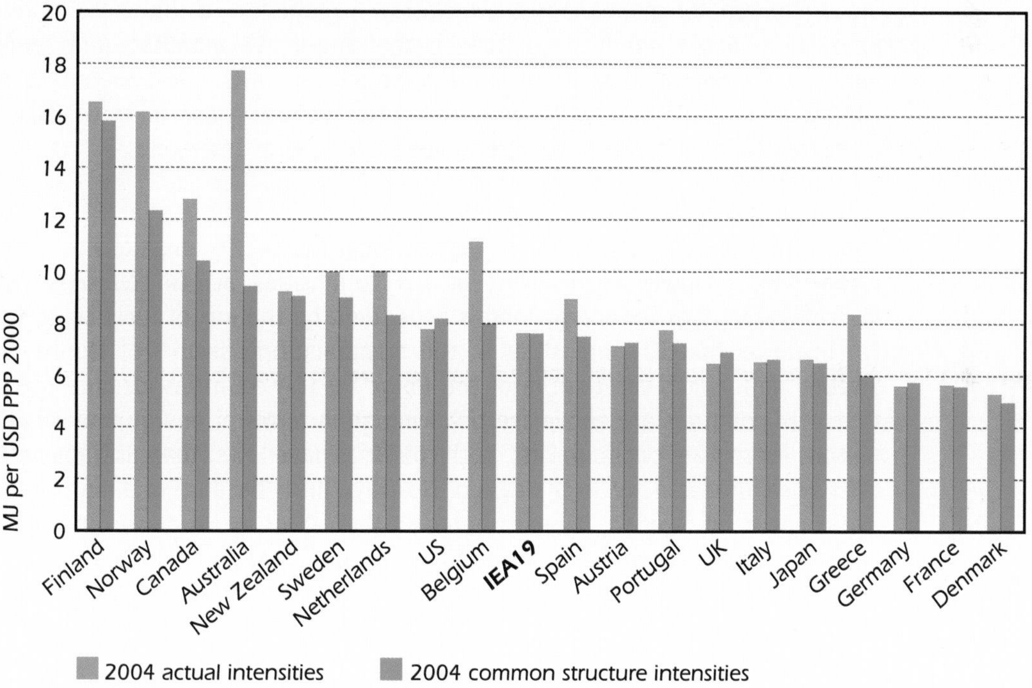

■ 2004 actual intensities    ■ 2004 common structure intensities

The results are quite striking for some countries, even though they do not account for detailed structural differences within the sub-sectors of industry (*i.e.* different product mixes). For instance, this approach shows that Australia's very high energy intensity can be largely explained by the structure of its manufacturing industry, which has a high share of very energy-intensive industries. If Australia's industry had the same structure as the average for the IEA19 countries – but kept its actual level of energy intensity in each sub-sector – the country's aggregate manufacturing energy intensity would be reduced by 47%. Similar, but less dramatic, results are also observed for Belgium, Canada, Greece, the Netherlands, Norway and Spain.

A similar result is obtained if manufacturing energy intensity is calculated using value-added at market exchange rates (see Figure B.2 in Annex B).

In most IEA countries, manufacturing energy use is concentrated in a few energy-intensive sub-sectors. The most energy-intensive sub-sector in the IEA19 is the production of primary metals. In 2004, energy intensity in this sub-sector was more than ten times higher than the category of "other manufacturing". Yet output from the primary metals sub-sector constituted only 4% of total manufacturing value-added, compared to more than 29% from other manufacturing. Taken together, the four most energy-intensive sub-sectors (primary metals, non-metallic minerals, chemicals, and paper and pulp) accounted for almost 70% of IEA19 manufacturing energy use, despite representing only 29% of total value-added output.

Clearly, changes in output shares can have large impacts on manufacturing energy use. It also follows that the larger the difference in energy intensities among different sub-sectors, the greater the impact such shifts in output might have on energy use. Consider the consequences when value-added from "other manufacturing" grows faster than that from primary metals. Each dollar generated in "other manufacturing" requires one-tenth (or less) of the energy input required for metals. Thus, even a small reduction in the share of metal production would significantly reduce the aggregate manufacturing energy intensity. Examining how these kinds of structural shifts affect manufacturing energy use is an important focus for the next section. These shifts will also be compared to impacts from changes in sub-sector energy intensities.

Turning to the intensities of $CO_2$ emissions, Figure 3.13 shows that all countries, again with the exception of Spain, showed a decline in $CO_2$ emissions per unit of value-added.[5] However, the level of $CO_2$ intensities across countries varies markedly from that for energy, and is dependent on the mix of fuels being used. For instance, the $CO_2$ intensities of the manufacturing sectors of Norway and Sweden lie towards the bottom of the range for IEA countries, despite the fact that both countries are above the IEA average in terms of energy intensity. The reasons for this are two-fold. First, both countries have lower than average carbon intensities for electricity. In the case of Norway, this is because of a high share of hydro power; in Sweden, it is attributed to the use of hydro power and nuclear energy. Second, the two countries also have a significant paper and pulp industry, which uses substantial amounts of biomass – an energy source that produces no net $CO_2$ emissions.

Conversely, countries such as Greece, the United Kingdom and the United States rank much higher on $CO_2$ emissions intensity than they do for energy intensity. This is largely because of high emissions from electricity generation that relies on substantial use of fossil fuel (and of coal, in particular).

5. Figure B.3 in Annex B shows $CO_2$ emissions intensity calculated using market exchange rates instead of purchasing power parity. This shows the same trend in intensities for all countries, although it changes the relative position of some countries.

**Figure 3.13 ▶** *CO$_2$ Emissions per Total Manufacturing Value-added*

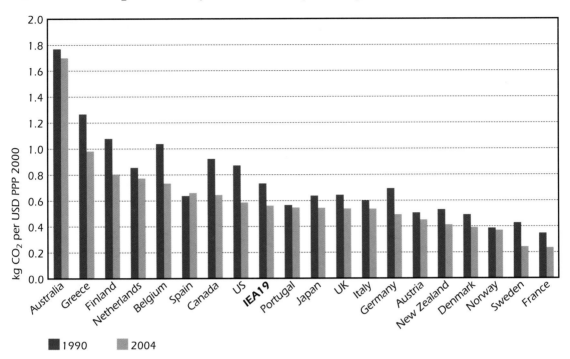

■ 1990    ■ 2004

# Impact of Changes in Structure and Intensity

Over the period 1990 to 2004, the sub-sector energy intensities for the IEA19 show significant declines for the most energy-intensive industries (primary metals, non-metallic minerals, chemicals, and paper and pulp). Of these, chemicals saw the largest reduction in energy intensity, falling by almost 30% between 1990 and 2004, with a particularly significant decline since 2000 (Figure 3.14). Some of the decline in energy intensity can be attributed to changes in the product mix within the industry itself, including a shift from basic chemicals (such as petrochemicals) to high value-added, low energy-intensity products (such as pharmaceuticals). Statistics from the OECD National Accounts suggest that in IEA countries pharmaceuticals increased its share in total chemicals production from around 30% in 1990 to almost 45% by 2004.

The energy intensities of non-metallic minerals and paper and pulp declined, by 16% and 13% respectively. In the case of non-metallic minerals, cement dominates the sub-sector. An increased share of cement production from more energy-efficient dry kilns helped drive the downward trend in energy intensity. In contrast, changes in paper and pulp can be attributed to increased recycling, updating of older plants, and to increased attention on energy management. For primary metals (*e.g.* iron, steel and aluminium), the decline in intensity was 8% over the same period, with an important contributions from increased recycling of scrap metal and some upgrading of existing plants.

The picture is more mixed for the less intensive sub-sectors of manufacturing. The most dramatic decline – a remarkable 46% – was in the energy intensity of metal products, which includes the manufacture of high-tech electrical goods. This sub-sector has seen very rapid growth in value-added since 1990 without a substantial increase in energy use. The energy intensity of other manufacturing declined by only 3%. In contrast to the general downward trend, food, beverages and tobacco experienced a slight increase in energy intensity of 4%, partly as a result of structural changes that included an increased share of heavily processed foods.

**Figure 3.14** ▶ *Evolution of Sub-sector Energy Intensities, IEA19*

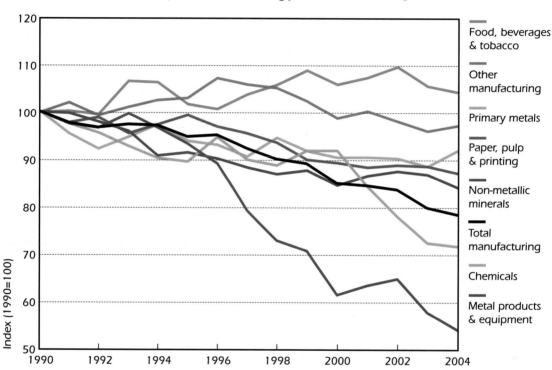

Figure 3.15 shows the average annual percentage change in actual manufacturing energy use and value-added (activity) for the IEA19, covering various time periods between 1990 and 2004. In all periods, the growth in energy use was much less than the growth in output. It is interesting to examine whether – and to what extent – these developments were driven by improvements in energy efficiency and/or by structural changes. The answer can be calculated by applying a decomposition approach (explained in Annex A). In Figure 3.15, the third bar represents the impact of structural changes; the fourth bar shows the effect of changes in energy intensities, adjusted for these structural changes (which are used as a proxy for energy efficiency).

In the first half of the 1990s, value-added from manufacturing industry increased by an average of 1.6% per year; energy use grew by only 0.2% per year. As a consequence, aggregate energy intensity fell by 1.3% per year. Because the manufacturing structure as a whole became less energy intensive, the structure-adjusted intensities fell by less, about 1.1% per year on average.

**Figure 3.15** ▶ *Factors Affecting Manufacturing Energy Use, IEA19*

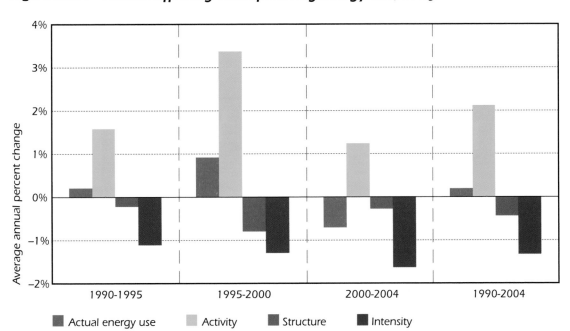

**3**

A similar pattern emerges in the second half of the 1990s. However, during this period the rate of growth in output more than doubled to 3.4% per year on average and energy use also grew more quickly. Thus, energy intensity declined at a faster rate of 2.2% per year. During this period, structural change (particularly a rapid increase in output from lower intensity sub-sectors such as metal products) was more significant in helping reduce the overall intensity of manufacturing. Even so, the structure-adjusted intensities fell by 1.3% per year on average. The same underlying pattern continued after 2000, except that growth in manufacturing output fell back to 1.2% and energy use actually fell by 0.7% per year on average, due to ongoing structural changes and faster sub-sectoral intensity declines.

Therefore, over the whole 1990 – 2004 period, the rate of decline in the structure-adjusted energy intensities averaged 1.3% per year and was the most important factor restraining growth in manufacturing energy use. However, these reductions in energy intensity were significantly lower than in earlier decades. From 1973 to 1990, intensity reduction averaged 2.8% per year (see Annex C). Structural changes also helped to dampen energy demand. This effect was most pronounced during the late 1990s – a period of rapid growth in output, stimulated by the less intensive sub-sectors.

Looking at the country level, Figure 3.16 examines the factors affecting aggregate energy intensities (Energy/Output). This reveals that structural changes had an important impact on manufacturing energy use in Finland, Japan, Norway and Sweden. In all of these countries, structural changes reduced energy use by 1.0% to 1.5% per year on average. In the case of Sweden, this reduction was augmented by a sharp decline in the structure-adjusted intensity, leading to a reduction in aggregate energy intensity of 3.6% per year. In contrast, structure-adjusted intensities increased in Japan and Norway, but because industry moved towards a less energy-intensive structure, there was a decrease in aggregate energy intensities.

**Figure 3.16** ▶ *Decomposition of Changes in Manufacturing Energy Intensity, 1990 – 2004*

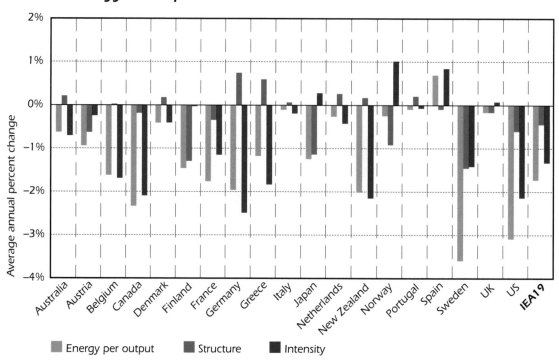

Although the impact of structural change was smaller in other IEA countries, understanding its effect is important to explaining developments in aggregate intensity. For instance, over the period 1990 to 2004, the United States showed a much larger reduction in aggregate energy intensities than Canada (3.1% as against 2.3% per year). However, once structural changes are taken into account, the intensity effect in both countries was remarkably similar, showing a reduction of 2.1% per year. This demonstrates that using only aggregated data can be misleading when comparing developments in energy efficiency across countries.

Figure 3.17 presents developments in $CO_2$ emissions from manufacturing for a number of time periods and compares them with trends in energy use. Overall, between 1990 and 2004, $CO_2$ emissions increased by 1% – less than the 3% increase in energy use. Two key factors affect the relationship between energy use and $CO_2$ emissions: the mix of energy commodities and the carbon intensity of the individual commodities. Across all time periods, changes in carbon intensity have reduced $CO_2$ emissions, largely as a result of a reduction in the $CO_2$ intensity of electricity. Between 1990 and 1995, changes in the energy mix – particularly a decline in the share of coal – also helped to reduce $CO_2$ emissions. As a result, even though energy use was on the rise, $CO_2$ emissions fell by 0.2% per year. In contrast, a rising share of electricity within the energy mix increased $CO_2$ emissions between 1995 and 2000. This counteracted the reduction in carbon intensity such that $CO_2$ emissions increased in line with energy use. A similar pattern was repeated after 2000. However, as energy use was decreasing in this period, $CO_2$ emissions also declined.

**Figure 3.17** ▶ *Decomposition of Changes in Manufacturing $CO_2$ Emissions, IEA19*

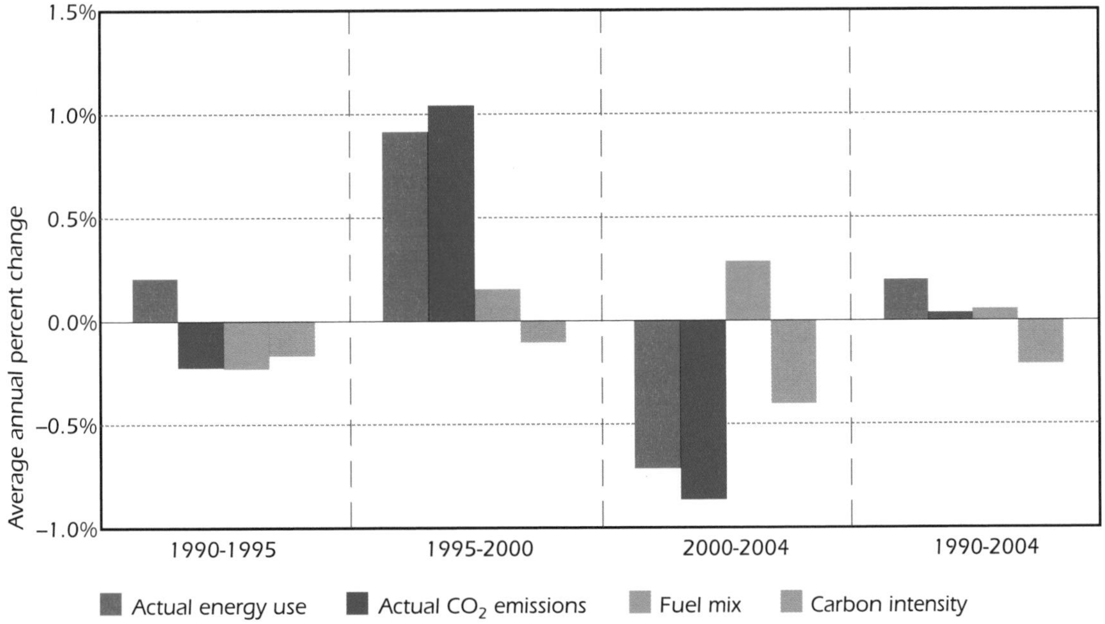

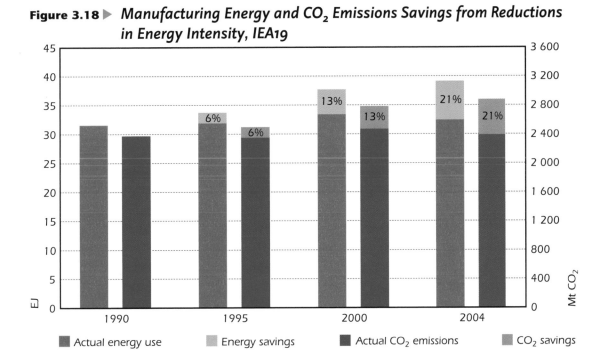

## Energy and $CO_2$ Savings

Between 1990 and 2004, changes in the structure of manufacturing and reduced energy intensities (i.e. improved energy efficiency) resulted in energy savings for the IEA19 countries. Almost 75% of these energy savings can be attributed to a decline in energy intensities. In 2004, energy savings from reduced energy intensities were 6.7 EJ or 21% of 2004 final energy use (Figure 3.18). The lower levels of energy intensity also reduced $CO_2$ emissions by 490 Mt in 2004.

**Figure 3.18** ▶ *Manufacturing Energy and $CO_2$ Emissions Savings from Reductions in Energy Intensity, IEA19*

Figure 3.19 provides the breakdown, by sub-sector, of energy savings deriving from changes in structure-adjusted energy intensities. The largest individual contribution came from chemicals, followed by paper and pulp, primary metals, and non-metallic minerals. Together these four energy-intensive industries accounted for more than two-thirds of total savings.

**Figure 3.19** ▶ *Breakdown of Energy Savings by Manufacturing Sub-sector, IEA19, 2004*

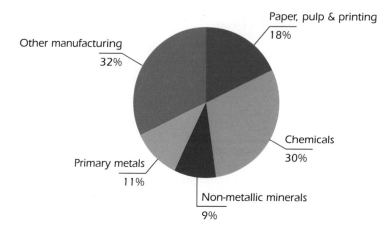

## Summary

Examination of the manufacturing sector of 19 IEA countries over the period from 1990 to 2004 reveals the following key findings:

▷  Total final energy use in this sector increased by only 3%, despite the fact that manufacturing output (as measured by value-added) increased by 31% over the same period. $CO_2$ emissions showed an even greater decoupling from economic growth, with an increase in emissions of just 1%.

▷  The majority of energy consumption in manufacturing is used to produce raw materials such as paper and pulp, chemicals, non-metallic minerals and primary metals. In 1990, the share of these sub-sectors was 72%; this decreased slightly to 69% by 2004.

▷  Natural gas continues to be the most important fuel used in the manufacturing sector; its share of final energy consumption reached 29% in 2004, up from 27% in 1990. Electricity use rose even more quickly, increasing its share from 24% to 27%. In contrast, the consumption of both oil and coal declined, continuing a downward trend that dates back to the 1970s. These trends can largely be explained by changes in relative fuel prices, shifts in industry structure and processes, and environmental legislation that favours the use of cleaner fuels.

▷  As a result of changes in the final energy mix, $CO_2$ emissions from manufacturing grew less quickly than final energy consumption. In many countries, the shift away from coal and oil – and towards natural gas – led to a decline in the $CO_2$ intensity of fossil fuel use. However, the increasing share of electricity partly offset these benefits. Despite improvements in the overall efficiency with which electricity is produced, electricity generation is still largely based on fossil fuels in many IEA countries.

▷ Final energy use is increasing much more slowly than manufacturing output. Thus, there was a 22% decline in overall energy intensity of manufacturing (as measured by final energy use per unit of value-added). All countries included in the analysis have shown reductions in their energy intensity with the exception of Spain, which recorded an increase.

▷ All of the most energy-intensive sub-sectors (primary metals, non-metallic minerals, chemicals, and paper and pulp) experienced declines in their energy intensities, with chemicals achieving the largest reduction of almost 30%. The picture is more mixed for the less intensive industries of manufacturing, with some sub-sectors increasing their intensity while others fell.

▷ The structure of the manufacturing sector became slightly less energy intensive. Yet, for most countries, reductions in the energy intensity of individual sub-sectors had more impact than structural changes on reducing the aggregate energy intensity of manufacturing. However, structural changes had an important impact on manufacturing energy use in Finland, Japan, Norway and Sweden.

▷ The rate of decline in the structure-adjusted energy intensities averaged 1.3% per year for the IEA19. However, this reduction was significantly lower than in earlier decades; the intensity reduction between 1973 and 1990 averaged 2.8% per year.

▷ Without these reductions in the structure-adjusted intensities since 1990, energy use in manufacturing would have been 21% higher in 2004. This represents an annual energy saving of 6.7 EJ and 490 Mt of avoided $CO_2$ emissions. More than two-thirds of the savings came from four energy-intensive industries: chemicals, paper and pulp, primary metals and non-metallic minerals.

**3**

# HOUSEHOLDS

## Scope

The household sector includes those activities related to private dwellings. It covers all energy-using activities in apartments and houses, including space and water heating, cooking, lighting and the use of appliances. It does not include personal transport, which is covered in a separate chapter.

## *Highlights*

Between 1990 and 2004, the overall energy efficiency of households in a group of 15 IEA countries improved by 0.7% per year. Without the energy savings resulting from these improvements, household energy consumption in the IEA15 would have been 11% higher in 2004 (Figure 4.1). This represents an annual energy saving of 2.7 EJ in 2004, which is equivalent to 190 Mt of avoided $CO_2$ emissions.

However, despite these savings, total final energy use (corrected for yearly climate variations) in households increased by 14% between 1990 and 2004. The increase was driven largely by a rapid rise in electricity use for appliances. The rate of energy efficiency improvement was much lower than in previous decades; an improvement of 2.0% per year was seen between 1973 and 1990.

**Figure 4.1** ▶ *Household Energy Savings from Improvements in Energy Efficiency, IEA15*

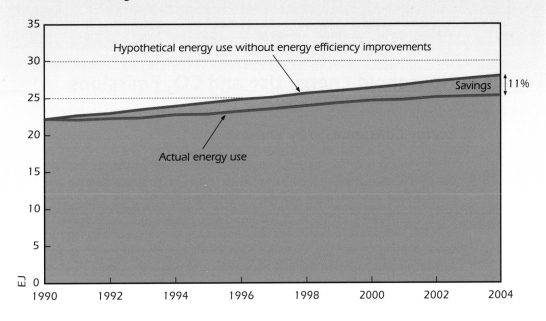

# Overview of Trends in the Household Sector

Between 1990 and 2004, total final energy use in the household sector (corrected for yearly climate variations[1]) rose by 14% in 15 IEA countries.[2] During the same period, population levels in these countries increased by 10% (Figure 4.2). Final energy use grew more quickly than population, revealing that the per capita energy use in households increased by 4%. This was partly due to increases in the number of households, which grew at a faster rate than population in most IEA countries. During the same period, $CO_2$ emissions grew by 15%, reflecting a slight increase in the carbon intensity of energy use in the sector.

**Figure 4.2** ▶ *Overview of Key Trends in the Household sector, IEA15*

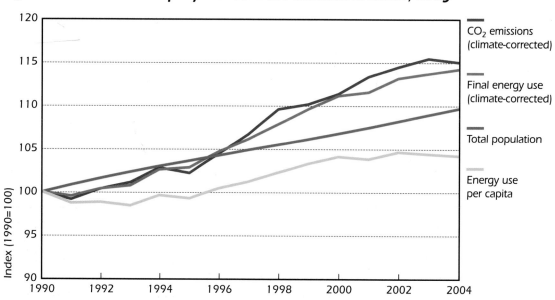

$CO_2$ emissions (climate-corrected)

Final energy use (climate-corrected)

Total population

Energy use per capita

# Trends in Household Energy Use and $CO_2$ Emissions

## Energy Consumption by End-use

Space heating is by far the most important energy user in the residential sector of the IEA15, accounting for 14 EJ in 2004 (Figure 4.3). Space heating energy use, corrected for yearly climate variations, increased by 5% since 1990. However, its share of energy consumption in the sector actually fell from 59% in 1990 to 54% in 2004. This reflects a significant reduction in the per capita energy requirement for space heating, driven by a combination of higher efficiencies of space heating equipment and improved thermal performance of new and existing dwellings.

---

1. Throughout this publication, energy use and $CO_2$ emissions for space heating in households have been corrected for yearly climate variations, except where explicitly noted.

2. The 15 IEA countries included in the analysis of the household sector are Austria, Canada, Denmark, Finland, France, Germany, Italy, Japan, the Netherlands, New Zealand, Norway, Spain, Sweden, the United Kingdom and the United States. These countries account for 88% of total household energy use in all IEA countries.

**Figure 4.3** ▶ *Household Energy Use by End-use, IEA15*

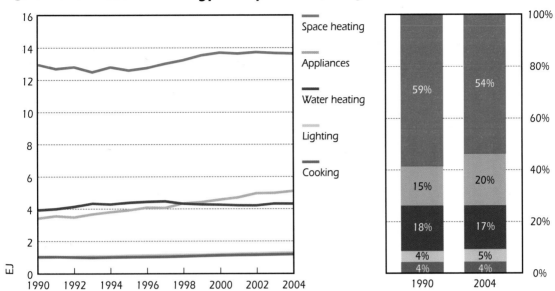

Space heating fuel shares vary significantly from country to country (Figure 4.4). In the IEA15 countries as a whole – and in many individual countries – natural gas is the main fuel for space heating, followed by oil (including oil products such as liquefied petroleum gas). In most countries, the share of both oil and renewables has fallen, and has been replaced by natural gas or electricity. In 2004, natural gas accounted for 52% of fuel use for space heating, followed by oil (25%) and renewables (10%). However, individual countries have important differences. In Japan, oil remains the dominant fuel for space heating. Electricity is important in space heating for New Zealand, Norway and Sweden, but represents only 9% for the IEA15 overall. In Denmark, Finland and Sweden, district heating represents the most important energy commodity for space heating. In the IEA15 countries as a whole, it represents only 4% of total energy use. The use of coal for space heating is now almost zero.

The most rapidly growing household demand for energy is from appliances, with consumption increasing by 50% from 1990 to 2004. In the late 1990s, appliances overtook water heating as the second most energy consuming category, accounting for 20% of total household energy consumption. In contrast, the share of water heating fell to 17% in 2004. The remaining end-uses – *i.e.* lighting and cooking – each account for around 4 to 5% of final energy use.

In all countries, with the exception of Norway and the United Kingdom, growth in energy demand from appliances out-paced growth in residential energy consumption on the whole. Finland, France and the United States are particularly noteworthy in that appliances energy demand rose by more than 70% between 1990 and 2004. This strong growth in overall energy consumption in appliances can largely be attributed to a rapid increase in use of a wide range of small electrical appliances and, in some cases, air conditioning (see Box 4.1). In Norway and the United Kingdom, energy use for space heating increased most quickly.

**Figure 4.4** ▶ *Share of Space Heating by Fuel*

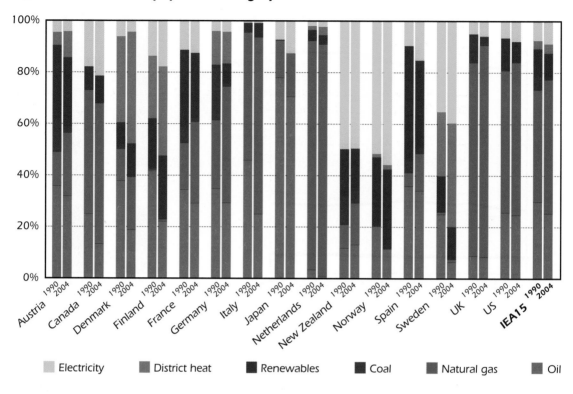

## Energy Consumption by Fuel

Electricity and natural gas consumption for households showed the largest increases, rising 35% and 23%, respectively (Figure 4.5). Strong growth in electricity consumption was largely driven by increases in the ownership and use of electrical appliances and, in

**Figure 4.5** ▶ *Household Energy Use by Fuel, IEA15*

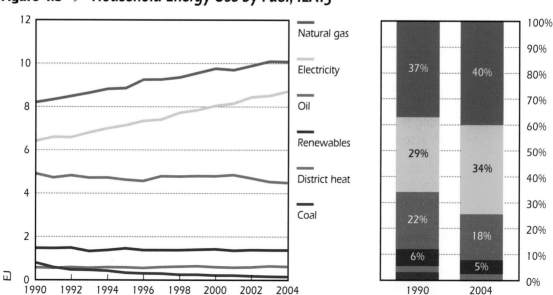

particular, a wide range of smaller appliances. The main drivers behind higher natural gas consumption include increased energy use for space heating and, to a lesser extent, fuel switching away from coal and oil. Between 1990 and 2004, the share of natural gas use rose from 37% to 40%; the share of electricity rose from 29% to 34%.

Coal's share declined to a negligible level of 0.5%, continuing a longer term downward trend. The importance of oil and renewables also declined. Oil consumption fell 8%; renewable energy consumption (mostly biomass) fell 7%. Changes in the energy mix reflect two factors: fuel substitution within the same end-use; and structural shifts among various household energy uses, i.e. stronger growth in one end-use compared to another. The reductions in coal, oil and renewables resulted, in part, from a switch to natural gas for space heating.

## $CO_2$ Emissions

The strong increase in electricity demand from appliances in the IEA15 countries was the main reason behind a 15% rise in residential $CO_2$ emissions between 1990 and 2004 (Figure 4.6). The increase in $CO_2$ was slightly higher than the rise in energy consumption, reflecting the increased share of electricity (see Box 2.1 in Chapter 2). $CO_2$ emissions from appliances, which rose by 44%, are rapidly approaching those from space heating as the largest source from households. Emissions from water heating, cooking and lighting are growing slowly; together, they account for 27% of total household $CO_2$ emissions.

**Figure 4.6** ▶ *Household $CO_2$ Emissions by End-use, IEA15*

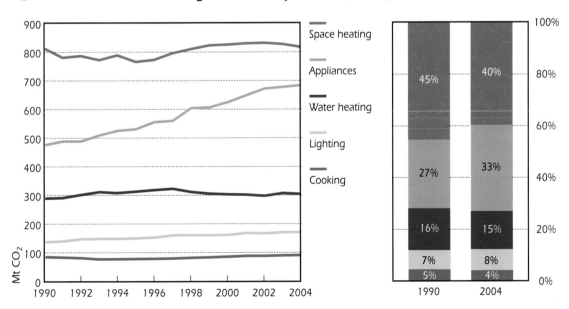

**Box 4.1**

## Electricity Use in Appliances

Electricity use for appliances in the residential sector of the IEA15 grew 48% from 1990 to 2004, driving the overall increase in household electricity demand. In 2004, electricity consumption in appliances was 57% of total household electricity use (Figure 4.7).

**Figure 4.7** ▶ *Household Electricity Demand by End-Use and the Role of Appliances, IEA15*

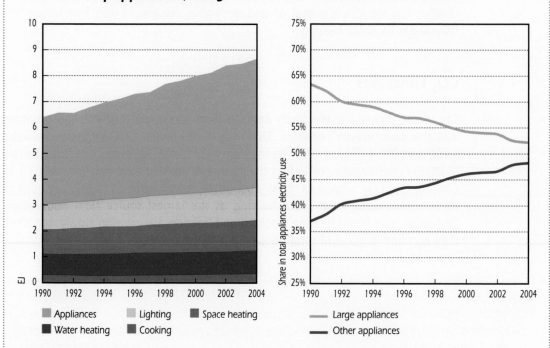

Appliances ◼ Lighting ◼ Space heating ◼ Water heating ◼ Cooking

— Large appliances — Other appliances

Information on the breakdown of appliances energy use is not available on a consistent basis across the IEA; several countries have no data. However, some data are available for large appliances such as refrigerators, freezers, televisions, washing machines and dishwashers. Many countries have already implemented minimum energy performance standards (MEPS) and labels for these appliances. The impacts of these energy efficiency programmes can be seen in data that show marked improvements in unit energy consumption. As an example, Figure 4.8 shows the average unit energy consumption (left) and total energy consumption (right) for these five large appliances in a group of 15 European countries (EU15). With the exception of televisions, all these appliances have shown a significant decrease in average unit energy consumption since 1990. In the case of refrigerators and freezers, the average unit energy consumption has declined even though the appliances themselves have become larger. For televisions, energy efficiency gains have been outstripped by the consumer trend towards larger screens, which use more energy. Thus, total energy consumption in the EU15 fell only in the case of refrigerators and washing machines. For other appliances, improved unit consumption has been more than offset by higher levels of ownership and use.

**Box 4.1** continued

**Figure 4.8** ▶ *Energy Consumption of Large Appliances, EU15*

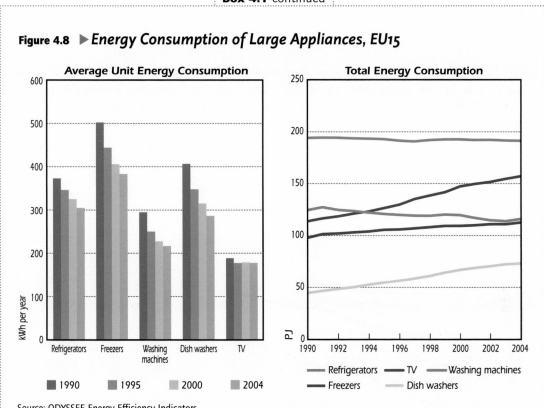

Source: ODYSSEE Energy Efficiency Indicators.

Currently, these five major appliances account for around 50% of household electricity consumption in appliances in the IEA15. However, this share is declining as the most rapid increase in appliance energy consumption comes from increasing ownership of a wide range of mostly small, miscellaneous appliances such as personal computers, mobiles phones, personal audio equipment and other home electronics. In some countries, air conditioning is also a key factor (see Box 4.2).

# Drivers of Energy Use in Households

Population levels play an important role in determining total energy use in the household sector. However, it is clear that other factors, such as household incomes and the cost of energy, are also important. For instance, higher incomes allow people to live in bigger apartments and houses, or perhaps have their own apartment or house instead of sharing with relatives and friends. As people grow richer, they also tend to purchase more and larger energy-consuming equipment. Higher energy costs are likely to restrain household energy consumption to some extent by encouraging more energy-efficient behaviour and purchases. However, energy costs continue to be a small proportion of household expenditure.

There is an obvious relationship between residential floor area per capita and income (Figure 4.9). As incomes rise, dwelling areas also tend to increase. For the IEA15 countries between 1990 and 2004, individual energy consumption per capita rose by 29%. During the same period, the average dwelling size per capita increased by 17%, to 49 square metres. This rise in average dwelling area has put upward pressure on per capita space heating and cooling needs.

**Figure 4.9** ▶ *House Area per Capita and Personal Consumption Expenditures, 1990 – 2004*

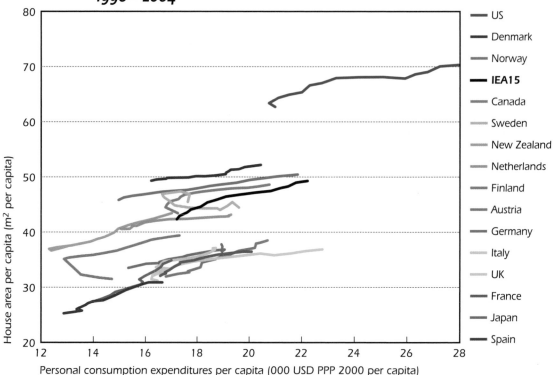

A large variation in the dwelling area per capita among the IEA15 countries is one of the factors explaining the differences in per capita space heating consumption. In 2004, the average person in the United States had almost 50% more dwelling area than the average for the IEA15. Part of this can be explained by differences in income levels. However, there are also significant differences in dwelling area per capita at a given income level. For instance, the United Kingdom has the second highest income per capita, yet also one of the lowest average house areas per capita. Many other factors affect house area per capita including land area per capita (which affects land, and therefore house, prices) and the level of urbanisation.

Changes in per capita dwelling area are related to the average dwelling size and the number of occupants in each household. Table 4.1 shows that in all IEA15 countries except Sweden (where occupancy rose very slightly), average household occupancies have fallen between 1990 and 2004. In 2004, the average number of people per household was 2.6, down from 2.8 in 1990. The reduction in the occupancy of each household puts upward pressure on demand for space conditioning in two ways. First, a larger number of dwellings are required to house

**Table 4.1** ▶ *Trends in Occupancy and Dwelling Area by Country*

| | Population per Occupied Dwelling | | | Average Dwelling Size (m²) | | |
|---|---|---|---|---|---|---|
| | 1990 | 2004 | % Change | 1990 | 2004 | % Change |
| Austria | 2.7 | 2.4 | -10 | 84 | 91 | 8 |
| Canada | 2.8 | 2.6 | -8 | 115 | 119 | 4 |
| Denmark | 2.2 | 2.1 | -3 | 107 | 110 | 2 |
| Finland | 2.4 | 2.1 | -12 | 75 | 78 | 4 |
| France | 2.7 | 2.5 | -7 | 86 | 91 | 5 |
| Germany | 2.5 | 2.3 | -7 | 82 | 86 | 5 |
| Italy | 2.9 | 2.6 | -9 | 93 | 96 | 3 |
| Japan | 3.2 | 2.7 | -15 | 86 | 93 | 8 |
| Netherlands | 2.6 | 2.4 | -6 | 104 | 104 | 0 |
| New Zealand | 2.9 | 2.8 | -5 | 100 | 115 | 16 |
| Norway | 2.4 | 2.3 | -6 | 110 | 114 | 3 |
| Spain | 3.5 | 3.0 | -14 | 86 | 91 | 5 |
| Sweden | 2.2 | 2.2 | 1 | 98 | 96 | -2 |
| United Kingdom | 2.6 | 2.4 | -7 | 80 | 88 | 10 |
| United States | 2.7 | 2.7 | -1 | 147 | 168 | 15 |
| IEA15 | 2.8 | 2.6 | -7 | 109 | 120 | 10 |

**4**

a given population. Second, the heating/cooling needs of a given dwelling are not necessarily reduced because fewer people are living in it. As a result, per capita space conditioning rises as occupancy falls.

The decline in the number of occupants per dwelling has been accompanied by an increase in the average size of dwellings. The combined effect of fewer inhabitants and larger average dwelling sizes led to a strong increase in residential floor area per capita (Figure 4.9). The United States, which has the highest income among the IEA15 countries, also has the largest homes (followed by Canada and New Zealand). Europe and Japan have the smallest average dwelling sizes, although homes in Scandinavia (except Finland) are larger than elsewhere in Europe.

The upward pressure on energy consumption from larger homes and lower occupancy levels has been offset to some degree by a decline in the space heating intensity. This is measured in terms of "useful energy" for space heating per square metre, with useful energy being calculated as final energy minus the losses estimated for boilers. To allow for comparisons across countries with different climates, the space heating intensity is divided by each country's yearly number heating degree-days (Figure 4.10).[3]

---

3. The concept of heating degree-days (HDD) is explained in more detail in Annex A.

**Figure 4.10 ▶ *Useful Space Heating Intensity in Households***

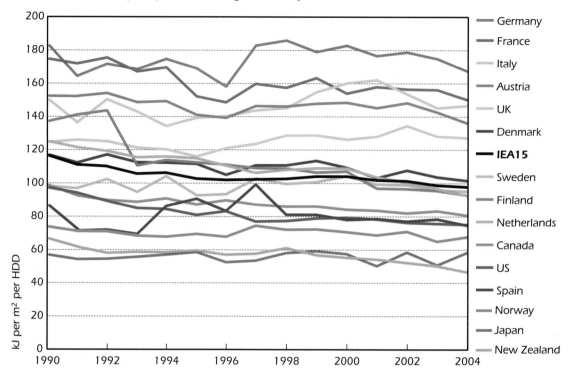

For the IEA15 as a whole, the useful energy per area heated declined by 16% between 1990 and 2004. Japan and the United Kingdom are the only countries in which energy intensities increased – and even then only slightly. This continues a longer term trend in both countries and reflects a situation in which income growth has led to higher heating comfort levels. In turn, this increased comfort level offset some of the savings from better insulation rates. Countries with relatively cold climates (*e.g.* Canada, Norway, Sweden and the United States) had relatively lower intensities. This is partly due to higher levels of insulation in some of these countries as well as a larger share of central heating, which is generally more energy efficient. France, Germany and Italy, which have more moderate winters, showed the highest levels of energy intensity. This could be the result of a greater percentage of poorly insulated older buildings.

Energy costs are another important factor to consider when examining trends in energy use. For the majority of IEA countries, the share of disposable incomes that households currently pay for energy is between 2% and 3% (Figure 4.11). Since 1990, the trend in many countries has been slightly downward, indicating that households are spending a smaller fraction of their income on energy. Notable exceptions to this trend are Denmark, France and Portugal, where the shares of incomes spent on energy have shown modest increases. Overall, the low and declining share of income spent on energy is unlikely to provide significant economic incentives for improved energy efficiency.

**Box 4.2**

## *Household Air Conditioning*

Consistent and comprehensive data on energy use in household air conditioning are not currently collected. However, air conditioning represents a rapidly growing end-use in many countries. Estimates by the IEA suggest that air conditioning now accounts for around 9% of household electricity consumption in IEA countries, substantially more than in 1990.

In the United States, the proportion of homes with air conditioning is now more than 80%, up from 64% in 1990. In Canada, air conditioning is still not as widespread, but growth rates have been similar. Australia has seen a dramatic rise in the penetration of air conditioning since 1999; more than 60% of homes are now cooled. Ownership and use of air conditioners is also growing in Europe, particularly in France, Greece, Italy and Spain. Japan has a long history of air conditioning; with ownership levels of 85%, the Japanese market seems close to saturation and growth is modest.

These increased air conditioning loads are having an impact on $CO_2$ emissions, as well as on security of supply and national infrastructure investment. In many countries, air conditioners are now a major contributor to summer peak electricity demand. Since 2000, air conditioning loads have been a factor in a number of blackouts and power shortages that occurred in IEA countries. They are also driving the need for substantial investment in electricity supply infrastructure.

Many IEA countries have introduced policy measures to improve the energy efficiency performance of air conditioning. Examples include the "Top Runner" programme in Japan (which sets industry targets based on the most energy-efficient products on the market) and standards set by the Department of Energy in the United States. However, the impact of more efficient individual air conditioners has often been outweighed by increased numbers and use of cooling equipment.

**4**

**Figure 4.11 ▶** *Share of Household Energy Expenditures in Total Personal Consumption Expenditures*

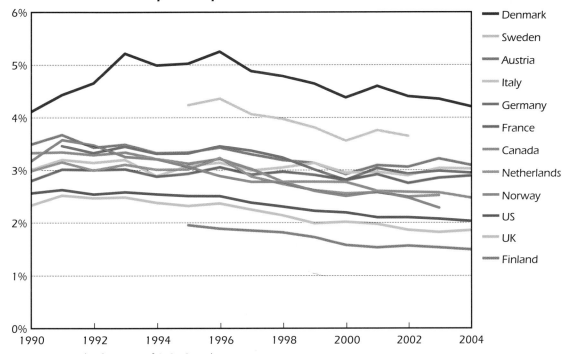

Source: OECD National Accounts of OECD Countries.

The percentage of total income spent on household energy depends on several factors: the price of purchased fuel; the mix of fuels used; and the level of residential energy demand per capita. The latter often depends, in turn, on the demand for space heating. Countries with high electricity and natural gas prices generally spend a larger portion of their income on residential energy than those countries that have lower energy prices.

For instance, households in Denmark spend the largest proportion of their income on energy, at 4.2% in 2004 (Figure 4.12). These high prices are partly a result of a green tax policy that places a $CO_2$ tax on household energy use. In Denmark, taxes represent 59% of the total electricity price; for natural gas, taxes represent 44% of the price. In other countries, tax represents a much lower percentage of the total price of both electricity and gas. In contrast, despite having a cold climate, Finnish households spend less than 2% of their income on energy, as both natural gas and electricity prices are low compared with the IEA average.

**Figure 4.12** ▶ *Household Energy Prices, 2004*

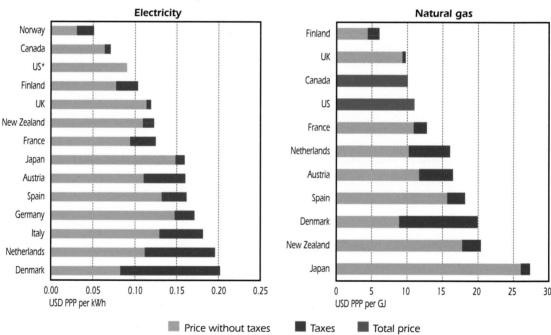

*Information on the tax component is not available.

Source: IEA/OECD *Energy Prices & Taxes*.

# Changes in per Capita Energy Use and $CO_2$ Emissions

The level of energy use per capita varies significantly across IEA countries, as illustrated by the square markers in Figure 4.13. In 2004, Canada had the highest consumption levels, followed closely by the United States and Finland. To a large extent, the high energy use per capita in Canada and Finland is due to their cold climates. In contrast, the United States has a milder climate. Thus, when energy use is normalised to the same climate conditions, the United States has the highest per capita consumption. At the other end of the scale, Japan, New Zealand and Spain had the lowest household energy use per capita in 2004. These countries all have relatively mild climates, but even after climate correction they still have the lowest per capita consumption.

**Figure 4.13** ▶ *Household Energy Use per Capita*

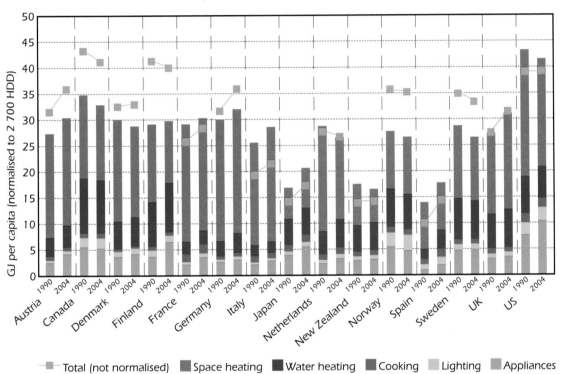

In general, once calculated on a climate-adjusted basis, variations in per capita energy use among countries are less dramatic. When their cold climates are taken into account, countries such as Canada, Finland, Norway and Sweden move from being at the high end of the consumption scale to near average levels (for the year 2004). However, even with a climate adjustment, on a per capita basis in 2004, the country with the highest consumption (the United States) used 2.5 times more energy than the country with the lowest consumption (New Zealand). It is noteworthy that this spread narrowed from a difference of three times in 1990.

Residential $CO_2$ emissions per capita – again corrected for yearly climate variations – for the IEA15 countries reveal a quite different pattern to those for energy (Figure 4.14). This is a result of differences in the fuel mix, particularly for electricity generation. The low emission levels in some countries are related to low or no emissions from electricity generation (see Box 2.1, Chapter 2). Conversely, the United States has relatively high emissions per kWh of electricity. When combined with high electricity consumption per capita, the United States has the highest per capita $CO_2$ emissions from residential electricity use within the IEA15.

**Figure 4.14** ▶ *Household CO$_2$ Emissions per Capita*

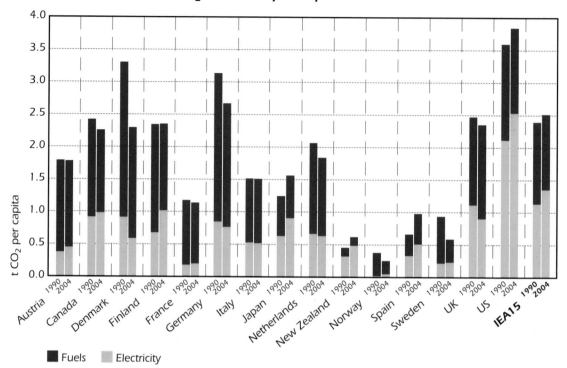

■ Fuels ▪ Electricity

Most of the fuel use in households is directed toward space heating. Thus, countries with relatively cold climates and a low share of electricity in the space heating mix can be expected to have high CO$_2$ emission levels from fuel use. Countries such as Denmark, Germany and the United Kingdom fall into this category. In Norway, the very high share of carbon-free hydro electricity, supplemented by biomass, results in low CO$_2$ emissions from space heating, despite a cold winter climate. Sweden is in a similar situation, although the electricity share of space heating is lower than in Norway. In addition, district heating from biomass is now common in Sweden, which helps to moderate emissions from fuel use. On the other hand, countries with low heating demands, such as New Zealand and Spain, often have low non-electricity emissions.

Countries that showed rising per capita emissions, most notably the United States and Japan, also showed higher emissions coming from electricity use. The United States generates significant shares of electricity from coal. Hence, it is more sensitive to rising emissions from electricity. In addition, both Japan and the United States experienced some of the strongest increases in energy consumption for appliances among the IEA15 countries. Finland and France also reported high energy demand from appliances, but this did not lead to higher emissions. Finland has a high share of nuclear and hydro electricity, with a much smaller percentage of coal-fired electricity. France's electricity mix is dominated by CO$_2$ emissions-free nuclear power.

## Impact of Changes in Structure and Intensity

Figure 4.15 summarises the impact that changes in activity, structure and end-use intensities have had on total residential energy use in the IEA15 countries. The activity component reflects population growth. Structural changes include three key factors: dwelling area per capita (for space heating and lighting); appliance ownership per capita; and household occupancy (for water heating and cooking). The intensity effect includes the impact of changes in all residential end-use intensities (Annex A explains the decomposition in more detail).

**Figure 4.15** ▶ *Factors Affecting Household Energy Use, IEA15*

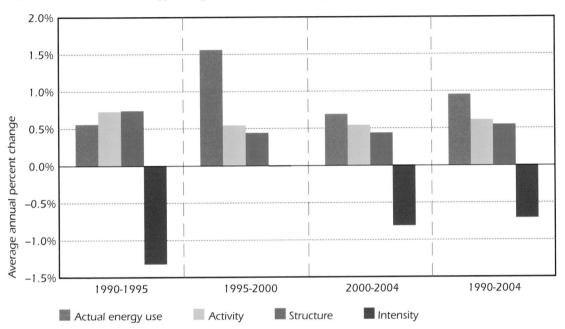

Residential energy consumption over the period between 1990 and 2004 grew at an average rate of almost 1.0% per year. Overall, the impact of activity and structure on energy use in the residential sector was relatively similar, with changes in structure having a slightly larger impact. Population growth caused energy to rise, on average, by 0.6% per year; changes in structure resulted in an annual increase of 0.5%. These two factors were partially offset by a 0.7% reduction per year in energy intensity.

However, examining the trends in more detail reveals significant differences across the time period. Between 1995 and 2000, energy consumption increased at 1.6% per year, more than double the rates before and after this period. The impact of activity and structure grew at relatively constant rates over the whole period. However, there were significant differences in the intensity effect, particularly during the late 1990s. Higher energy use between 1995 and 2000 was the result of upward pressure from activity and structural effects that were not offset by improvements in intensity (as were seen in the periods 1990 – 95 and 2000 – 04). This was as a result of the trend in energy intensity for space heating, which was virtually constant during the 1995 – 2000 period after showing significant improvement between 1990 and 1995 and, again, between 2000 and 2004.

As described earlier in this chapter, several factors affect space heating in households. The impacts of these effects are summarised in Figure 4.16. For most countries, larger dwelling sizes and fewer occupants per dwelling have tended to drive up energy demand for space heating. From 1990 to 2004, larger dwelling sizes in the IEA15 countries led to an annual increase of 0.6% in energy demand. Lower occupancy rates increased demand by 0.5% per year.

**Figure 4.16 ▶** *Decomposition of Changes in Space Heating per Capita, 1990 – 2004*

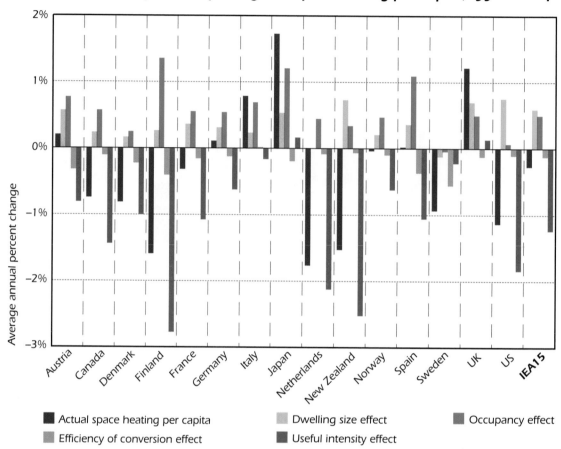

| ■ Actual space heating per capita | ■ Dwelling size effect | ■ Occupancy effect |
| ■ Efficiency of conversion effect | ■ Useful intensity effect | |

On average, higher demand for space heating in the IEA15 has been offset by two factors: efficiency gains derived from lower conversion losses (due to fuel switching only) and a significant reduction in the useful energy intensity for space heating. Lower end-use conversion losses led to energy demand reductions of 0.1% per year. The useful intensity effect caused a 1.2% reduction per year. The overall effect has been a small (0.3%) annual reduction in actual space heating per capita in the IEA15 countries from 1990 to 2004.

Most countries followed a similar pattern. The tendency for higher energy demand – caused by fewer occupants and larger homes – was offset by lower end-use conversion losses and, more importantly, a decline in the useful intensity of space heating. Japan and the United Kingdom were notable exceptions. For these countries, increased comfort levels (higher internal temperatures and, to a lesser extent, an increased penetration of central heating) resulted in higher energy intensity levels for space heating.

Overall $CO_2$ emissions from households grew by 1.0% per year in the IEA15 countries. This figure reflects the fact that higher activity and structural effects were partially offset by lower emissions deriving from reductions in energy and carbon intensity, as well as the effects of changes in the energy mix (Figure 4.17). The energy mix effect represents carbon savings from fuel substitution. Carbon intensity largely represents a cleaner electricity mix, as well as the use of less carbon-intensive coal or oil products.

The decomposition of individual country trends showed that, in most countries, structure was the most important factor in relation to increasing emissions. The exceptions were Canada and the United States, where activity was the leading contributor to higher emissions. New Zealand also showed a different trend in that higher carbon intensity was the primary contributor to increasing emissions. Six countries actually showed a reduction in total emissions from residential energy use, the most notable being Denmark, Norway and Sweden. Denmark pursued a cleaner electricity mix – replacing coal with wind power and natural gas – which resulted in reduced $CO_2$ emissions. In Norway, this decline is the result of fuel substitution from oil to carbon-free hydro electricity. In Sweden, the reduction derives from biomass-based district heating.

**Figure 4.17** ▶ *Decomposition of Changes in Household $CO_2$ Emissions, 1990 – 2004*

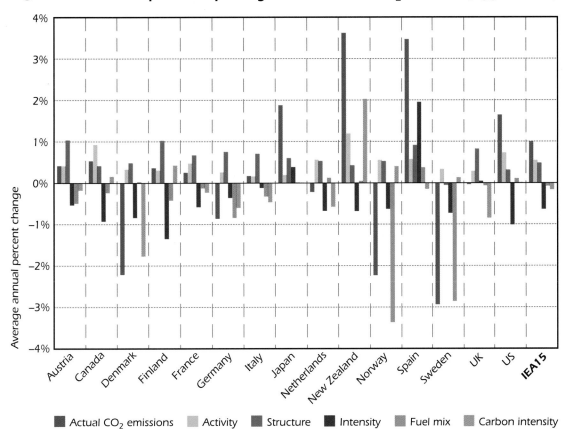

## Energy and $CO_2$ Savings

The reduction in energy intensity in the residential sector in the IEA15 countries between 1990 and 2004 led to energy savings of 11% or 2.7 EJ in 2004 (Figure 4.18). However, the average annual improvement in intensity over this period of 0.7% was considerably lower than in past decades; between 1973 and 1990, the improvement was 2.0% per year (see Annex C).

Over this same period, $CO_2$ savings showed a more modest decline of 9% or 190 Mt $CO_2$. The lower rate of decline was a result of changes to the end-use energy mix, which led to it becoming more carbon-intensive. This increased carbon intensity offset some of the carbon savings from the reduction in energy intensity.

**Figure 4.18** ▶ *Household Energy and CO₂ Emissions Savings from Reductions in Energy Intensity, IEA15*

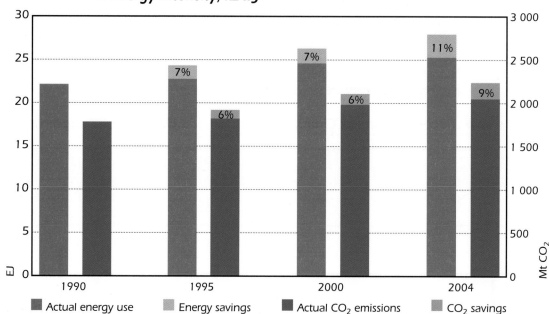

Legend: ■ Actual energy use   ■ Energy savings   ■ Actual CO₂ emissions   ■ CO₂ savings

## Summary

Examination of the household sector of 15 IEA countries over the period from 1990 to 2004 reveals the following key findings:

▷ Total final energy use in households, corrected for yearly climatic variations, grew by 14%. $CO_2$ emissions rose by 15%. This increase in energy use and $CO_2$ emissions was a result of population growth, coupled with changes to the structure of households and increasing appliance ownership.

▷ Increases in both electricity and natural gas consumption have been particularly significant, driven by higher appliance use and space heating requirements respectively. Since 1990, electricity consumption in households increased by 35%; natural gas use rose by 23%.

▷ At 54%, space heating remains the most significant use of energy by households, even though its share has declined slightly. In contrast, appliance energy consumption is growing rapidly and accounted for a share of 20% in 2004. Appliance energy use has overtaken water heating as the second most important energy-consuming category.

▷ The increased energy consumption in appliances caused $CO_2$ emissions from this end-use to increase by 44%. Most appliances use electricity which, in many IEA countries, is mainly produced from fossil fuels. Emissions from this end-use are rapidly catching those from space heating as the largest source from households.

▷ Per capita energy consumption increased by 4%. This can be attributed to increased appliance ownership, as well as to increases in the average size of dwellings and lower number of people per household. Improvements in the useful space heating intensity helped to offset some of these increases.

▷ The energy and $CO_2$ emissions increases from appliances are now being driven by a wide range of mostly small appliances, as well as by air conditioning in some countries. Policies such as minimum energy performance standards have had some impact in curbing the increase in energy consumption of large appliances. However, these large appliances now represent only 50% of total appliance energy consumption, and this share is falling.

▷ In most countries, the share of household incomes spent on energy is between 2% and 3%, a slight decline since 1990. There are substantial differences in energy prices from country to country, partly as a result of different tax levels. Overall, the low and declining share of incomes spent on energy is unlikely to provide significant economic incentives for improved energy efficiency.

▷ Between 1990 and 2004, the overall energy intensity of the household sector declined at an average rate of 0.7% per year. However, this improvement in intensity was considerably lower than the rate achieved between 1973 and 1990, at which time the improvement was 2.0% per year.

▷ Without the energy savings resulting from reductions in energy intensity, final energy consumption in the household sector would have been 11% higher in 2004. This represents an annual energy saving in 2004 of 2.7 EJ and 190 Mt of avoided $CO_2$ emissions.

4

# SERVICES

## Scope

The service sector includes activities related to trade, finance, real estate, public administration, health, education and commercial services.

## Highlights

Between 1990 and 2004, the overall energy efficiency of the service sector in a group of 19 IEA countries improved by 1.1% per year. Without the energy savings resulting from these improvements, energy consumption in this sector would have been 17% higher in 2004 (Figure 5.1). This represents an annual energy saving of 2.8 EJ in 2004, which is equivalent to 280 Mt of avoided $CO_2$ emissions.

Despite these savings, total final energy use in services increased by 26% between 1990 and 2004, driven predominantly by a 50% increase in electricity use. The rate of energy efficiency improvement was much lower than in previous decades; an improvement of 2.3% per year was seen between 1973 and 1990.

**5**

**Figure 5.1** ▶ *Service Energy Savings from Improvements in Energy Efficiency, IEA19*

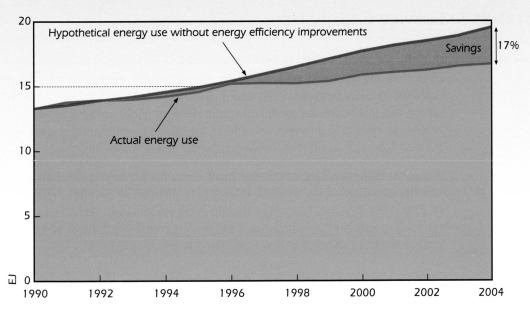

# Overview of Trends in the Service Sector

Between 1990 and 2004, the economic activity of the service sector, as measured by value-added output, showed a 45% increase in a group of 19 IEA countries[1] (Figure 5.2). During the same period, total final energy use increased by 26%. This decoupling of energy consumption and economic activity resulted in a 14% decline in energy intensity per unit of output from the service sector. $CO_2$ emissions grew by 29%, reflecting a slight increase in the carbon intensity of the final energy mix.

**Figure 5.2** ▶ *Overview of Key Trends in the Service Sector, IEA19*

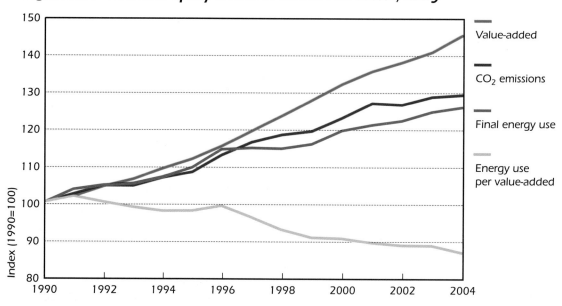

# Trends in Service Energy Use

Over the period from 1990 to 2004, total final energy use in services increased by 26%. This overall trend was driven by strong growth in electricity use, which increased by 50% between 1990 and 2004 (Figure 5.3). Electricity's share of the energy mix in the service sector rose from 43% in 1990 to 51% in 2004. The increase in electricity consumption reflects the growing importance of devices that use electricity such as lighting, office equipment and air conditioning.

In contrast to the rapid increase in electricity consumption, oil consumption fell 19% over the same period; its share of total service energy use declined from 21% in 1990 to 14% in 2004. Part of the decline in oil consumption was offset by higher natural gas consumption, which rose by 29%. The combined share of all other fuels (coal, renewables and district heat) decreased to 3% in 2004.

---

1. The 19 IEA countries included in the analysis of the service sector are Australia, Austria, Belgium, Canada, Denmark, Finland, France, Germany, Greece, Italy, Japan, the Netherlands, New Zealand, Norway, Portugal, Spain, Sweden, the United Kingdom and the United States. These countries account for 91% of total IEA energy consumption in services.

**Figure 5.3** ▶ *Service Energy Use by Fuel, IEA19*

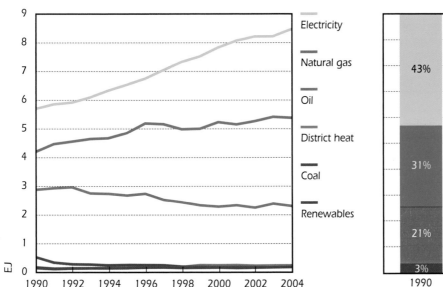
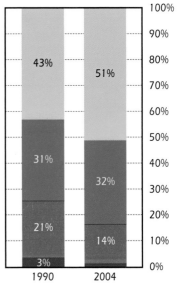

## Service Sector Energy Consumption by End-use

The availability of disaggregated end-use data for services is poor in most IEA countries. Some data exist for Canada and the United Kingdom, and are shown in Table 5.1. Combined with information from other sources (*e.g.* the EU ODYSSEE project, see Annex D), these data indicate that thermal uses of energy (space heating, water heating and cooking) account for just over half of the total energy use in services. However, other end-uses are also important, particularly those that consume electricity such as lighting and office equipment. Space cooling still has a relatively small share, but is growing rapidly in many countries.

**Table 5.1** ▶ *Percentage Share of Service Sector Energy Consumption by Energy Commodity and End-use, 2004*

|  | Space heating | | Space cooling | | Lighting | | Other | | Total | |
|---|---|---|---|---|---|---|---|---|---|---|
|  | Electricity | Other fuels | Electricity | Other fuels | Electricity | Other fuels | Electricity | Other fuels | Electricity | Other fuels |
| Canada | 4 | 36 | 7 | 0 | 13 | 0 | 29 | 11 | 53 | 47 |
| United Kingdom | 7 | 39 | 5 | 0 | 20 | 0 | 17 | 12 | 49 | 51 |

# Drivers of Energy Use in the Service Sector

The main factor affecting energy use in services is the level of economic activity in the sector. This is linked both to population (the number of people demanding the various services) and income levels (with higher incomes, people generally demand more services). In turn, higher economic activity leads to increases in the stock of buildings (with associated energy demands) and more people are employed in the sector (often requiring energy-using equipment to do their jobs).

Over the period from 1990 to 2004, the level of economic activity in the service sector – as measured by value-added output[2] – grew by 45% in the IEA19. In most IEA countries, the increase in output from services contributed to the bulk of the growth in total GDP. Comprehensive, disaggregated data for services are not available from OECD National Accounts. However, the available information suggests that strong growth was experienced across many sub-sectors of services including wholesale, retail, financial intermediation and other business activities. In contrast, growth in public sector activities appears to be lower. This is not surprising as the demand for public sector services (schools, hospitals, local services, etc.) tends to be closely linked to population levels, which showed only a 10% increase between 1990 and 2004 in the IEA19.

As a result of this increase in service sector output for the IEA19, the share of service sector output in total GDP also showed a modest increase: from 68% in 1990 to 71% in 2004. In fact, this growth in the contribution of services to total economic output extends a longer term trend that dates back to the early 1970s. Most countries have shown a similar trend, with service sector GDP rising slightly as a percentage of total GDP (Figure 5.4). This increase was largest for Germany and Japan at 8 percentage points, followed by a rise of 6 percentage points in Portugal and the United Kingdom.

**Figure 5.4** ▶ *Share of Service Sector Output in Total GDP*

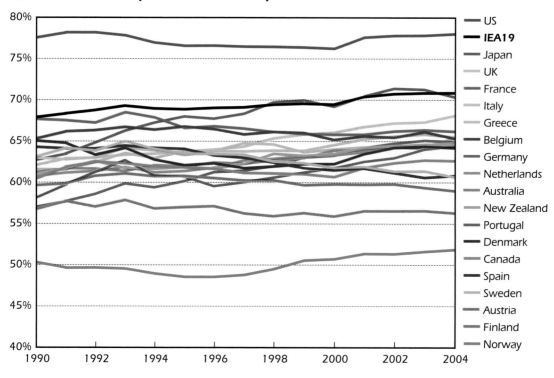

---

2. In this chapter, services value-added output is calculated in national currencies at 2000 constant prices and then converted to USD using the relevant purchasing power parities for the year 2000 (see Box 2.2 in Chapter 2). Selected charts have also been re-calculated using value-added output converted to USD at 2000 market exchange rates; these are shown in Annex B.

A number of countries showed slight declines in the share of services. In Finland and Sweden, large increases in manufacturing output (related to the growth of the cellular phone industry) reduced the share of the service sector in total GDP. In Norway, the share of services is significantly lower than the IEA average, primarily because the petroleum sector makes a particularly large contribution to the overall economy.

The expansion of the service sector has been accompanied by an increase in the stock of public and commercial buildings. This link between increases in service sector output and the total built area is important in that growth in floor area drives energy consumption for several key end-uses within buildings (*e.g.* space heating, space cooling and lighting).

Figure 5.5 shows how floor area per capita in the service sector evolved with respect to the growth in value-added per capita across IEA countries. Most countries show a similar relationship. For the ten IEA countries for which time-series floor area data are available, value-added in services rose by an average of 39% between 1990 and 2004, accompanied by a 24% increase in floor area. In Finland, France and Japan, the increase in floor area since 1990 has exceeded that of service output. In fact, France and Japan started with lower levels of floor space per unit of service sector value-added. Thus, this rapid growth has helped bring them both closer to the average for IEA countries.

**Figure 5.5** ▶ *Service Sector Floor Area per Capita and Value-added per Capita, 1990 – 2004*

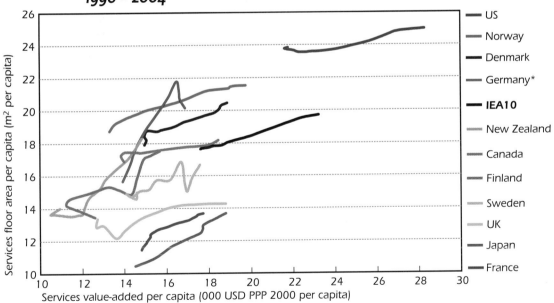

* Data for Germany start in 1995 and are not included in IEA10.

Growth in service sector value-added and floor area has, in turn, driven substantial increases in energy use, particularly in demand for electricity (Figure 5.6). Indeed, for the IEA19 countries, the growth in electricity between 1990 and 2004 outpaced the growth of service sector output.

**Figure 5.6** ▶ *Service Sector Electricity Use per Capita and Value-added per Capita,*
*1990 – 2004*

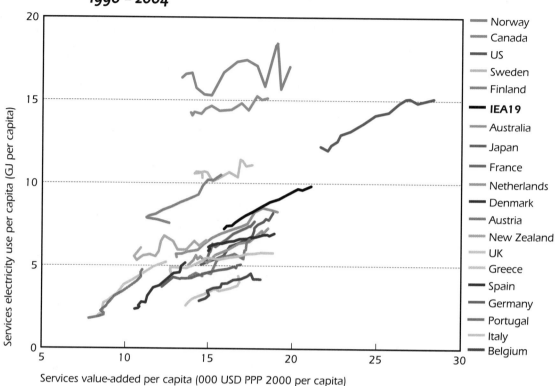

Strong growth in electricity use can be attributed to increased demand for space conditioning (heating and cooling) and lighting, as well as to the proliferation of a wide array of other electric end-uses. The use of office and network equipment grew quickly over this period as computers, photocopiers and printers became ubiquitous features in nearly all parts of the service economy.

Most of the countries depicted show strikingly similar relationships between electricity use and service value-added, with both rising significantly over this period. Greece, Portugal and Spain show the largest increases in electricity use relative to value-added output, partly as a result of an increased penetration of air conditioning. In contrast, Canada, New Zealand and Norway show a weaker link between increased electricity use and value-added. Both Canada and Norway already have high levels of per capita electricity consumption. In Canada, the high rates were linked to higher than average electricity use in equipment. Norway uses a great deal of electricity for space heating.[3]

# Changes in Aggregate Energy Intensity

In most countries, total final energy use in the service sector grew less rapidly than economic activity in the sector between 1990 and 2004. This resulted in a 14%

---

3. Figure B.4 in Annex B shows the same indicator, re-calculated with value-added output converted to USD using market exchanges rates. The same basic relationship between electricity use per capita versus value-added per capita is seen for each country. However, the country positions change horizontally with respect to one another. In particular, the positions of Japan and the United States shift substantially to the right, relative to other countries.

reduction in final energy intensity across the IEA19. As noted previously, economic activity (as measured by value-added) is an important driver for overall service energy use (particularly for electricity). However, other factors such as building stock (which grew more slowly than value-added) are also important for some end-uses such as space heating and lighting. Furthermore, because electricity has higher end-use conversion efficiency than other fuels, its rapid increase contributes to a reduction in final energy intensity. As a result, there was a partial decoupling of growth in total final energy use and economic activity.

The service sector in countries such as Canada, Germany, Norway, Sweden and the United States had the highest energy intensity in 1990; they also showed the greatest reductions between 1990 and 2004 (Figure 5.7). For many of these countries, higher intensities can be attributed primarily to colder climates that require greater space heating. The reverse trend was seen for Austria, Greece, Italy, Portugal and Spain, where energy intensity increased. With the exception of Austria, these countries also have some of the lowest energy intensities among the IEA19 countries. Overall, there appears to be a trend towards convergence in service energy intensity for IEA countries: the spread in intensities across countries fell by almost half between 1990 and 2004.[4]

**Figure 5.7** ▶ *Energy Use per Service Sector Value-added*

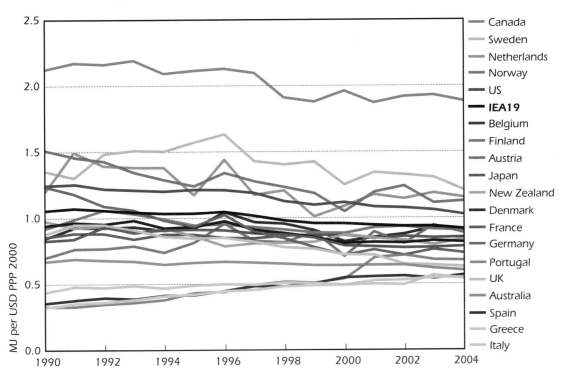

4. If services final energy intensities are re-calculated using value-added at market exchange rates (MER), a different picture emerges (Figure B.5, Annex B). The trends for individual countries are the same, but the position of countries relative to one another changes significantly. Using MER, the countries with the highest energy intensities are Canada, the Netherlands, New Zealand and Sweden; the lowest energy intensities are observed for Italy, Japan and the United Kingdom. When MER are used in the calculation, there is also much less convergence in energy intensities over time.

Interpreting these trends with respect to changes in energy efficiency is difficult, particularly in the absence of more detailed structural information. Different service sector activities can produce very different levels of economic output while consuming nearly the same amount of energy. For example, buildings in the finance sector can have the same final energy demand profile as buildings in the retail sector, yet generate significantly different levels of economic output.

An alternative measure of energy intensity for services is total final energy use per unit of floor area (Figure 5.8). In most countries, this measure of energy intensity fell at a lower rate than final energy use per unit of value-added. However, important differences are evident in the trends for electricity and other fuels. For the IEA10 group of countries (those for which floor area data are available), fuel use per unit of floor area fell by 12% between 1990 and 2004. In contrast, electricity use rose 17% per unit of floor area. As fuels are used mainly for space heating, this decline represents savings in space heating energy per unit of floor area. Higher electricity use in many countries can be explained by the growth of electricity end-uses such as lighting, office equipment and space cooling.

**Figure 5.8** ▶ *Electricity and Fuel Use per Unit of Floor Area*

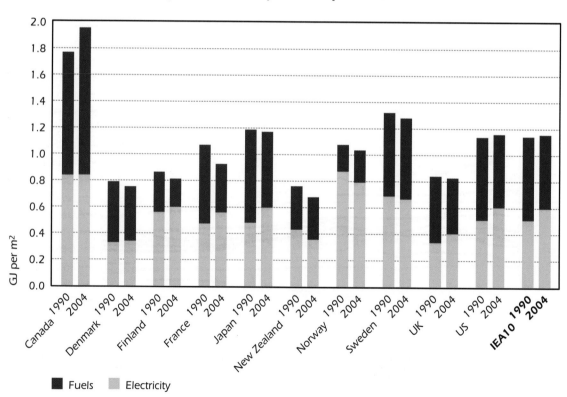

Canada and Norway showed the reverse trend – *i.e.* increasing fuel use and falling electricity use per floor area. In Canada, the increase was due to higher oil use for space heating. Historically, Norway has had high electricity consumption due to its use for space heating. In recent years, there has been a slight shift toward fuels for this end-use.

# Impact of Changes in Output and Intensity

Using the decomposition approach described in Annex A, Figure 5.9 summarises the impacts that changes in activity (value-added) and energy intensity (final energy per value-added) have had on total energy consumption in the service sector in the IEA19. Strong, sustained growth in service sector output, averaging 2.7% per year between 1990 and 2004, has driven growth in service sector energy use. However, the rate of growth in service energy use appears to be slowing. From 1990 to 1995, despite a period of relatively low growth in output, service energy use grew by 1.9% per year. Later in the 1990s, the growth in service output accelerated, but a sharp decline in energy intensity actually slowed the rate of increase in energy use. Since 2000, lower growth in service output and continued declines in energy intensity have further restrained the rise in energy use. Overall, between 1990 and 2004, energy intensity declined on average by 1.1% per year. However, this rate was substantially lower than the longer term trend between 1973 and 1990, when energy intensity declined by 2.3% per year (see Annex C).

**Figure 5.9** ▶ *Factors Affecting Service Sector Energy Use, IEA19*

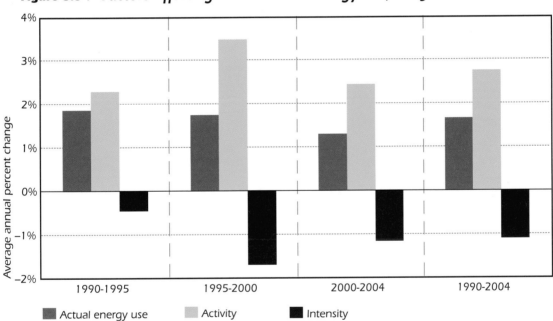

It is interesting to look more closely at the factors affecting trends in energy intensity in the service sector across various countries. Unfortunately, in the absence of comprehensive data, such an analysis has only been possible for some countries.

Figure 5.10 shows that across ten IEA countries, changes in three factors – energy use per floor area, floor area per employee and number of employees per unit of value-added (the inverse of labour productivity) – have interacted to reduce energy intensities (energy use per value-added). Improvements in labour productivity (*i.e.* reductions in the number of employees per unit of value-added) have occurred in all countries. For most countries, this has been the most important factor in reducing energy intensities in

**Figure 5.10** ▶ *Decomposition of Changes in Service Sector Energy Intensity, 1990 – 2004*

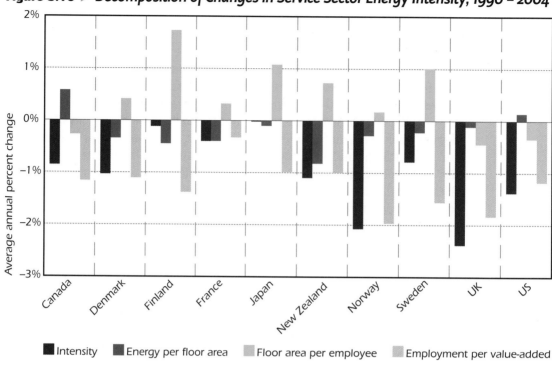

■ Intensity   ■ Energy per floor area   ■ Floor area per employee   ■ Employment per value-added

services, followed by reductions in energy use per unit of floor area. In contrast, a significant number of countries appear to have seen an increase in floor area per employee, which placed upward pressures on energy intensities.

Trends in $CO_2$ emissions in the service sector differ from those seen for energy for several reasons (Figure 5.11). Overall, $CO_2$ emissions grew by 29% between 1990

**Figure 5.11** ▶ *Decomposition of Changes in Service $CO_2$ Emissions, IEA19*

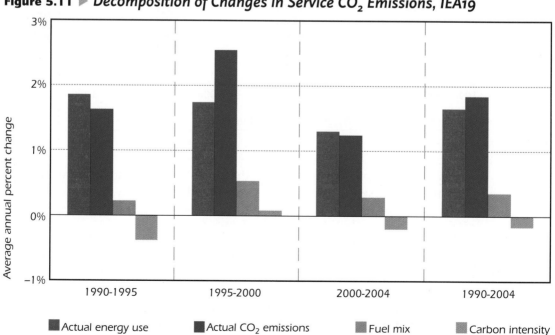

■ Actual energy use   ■ Actual $CO_2$ emissions   ■ Fuel mix   ■ Carbon intensity

and 2004, a larger increase than seen for energy use. Two factors affect the relationship between energy and $CO_2$ emissions: the mix of fuels used and the carbon intensity of the individual fuels. Across all time periods examined, changes in the energy mix have led to increased emissions of $CO_2$. This can be largely attributed to the increased share of electricity, which is still predominantly produced from fossil fuels in many IEA countries. The other factor, carbon intensity, depends largely on the fuel mix and the efficiency of electricity production. Overall, between 1990 and 2004, changes in carbon intensity have tended to decrease emissions. However, during the period 1995 to 2000, carbon intensity augmented the energy mix effect. As a result, $CO_2$ emissions increased much more quickly than final energy use.

# Energy and $CO_2$ Savings

Final energy use in the service sector grew 26% from 13.2 EJ to 16.6 EJ between 1990 and 2004 in the IEA19 countries. At the same time, a decline in energy intensity (energy per value-added) led to significant energy and $CO_2$ savings (Figure 5.12). In 2004, energy savings totalled 2.8 EJ, or 17%, of final energy use in the service sector. Energy intensity improvements also had a significant impact on $CO_2$ emissions, with a saving in 2004 of 280 Mt $CO_2$ (17% of actual $CO_2$ emissions in 2004).

**Figure 5.12** ▶ *Service Energy and $CO_2$ Emissions Savings from Reductions in Energy Intensity, IEA19*

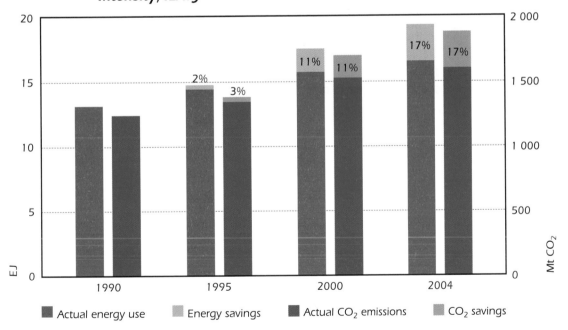

Legend: ■ Actual energy use ■ Energy savings ■ Actual $CO_2$ emissions ■ $CO_2$ savings

# Summary

Despite limited available data, examination of the service sector of 19 IEA countries over the period from 1990 to 2004 reveals the following key findings:

▷   Total final energy consumption increased by 26%; $CO_2$ emissions rose 29%. These increases were driven by rising output from the sector (as measured by value-added), which increased by 45%.

▷   The increase in electricity consumption has been particularly strong, reflecting the growing importance of electricity-using devices such as lighting, office equipment and air conditioning. Since 1990, electricity use has grown by 50%. With a share of 51% in 2004, electricity now dominates the service sector energy mix.

▷   The increasing share of electricity in the final energy mix meant that $CO_2$ emissions increased more quickly than final energy use. Despite improvements in the overall efficiency with which electricity is produced, electricity production is still largely fossil-fuel based in many IEA countries.

▷   Energy use is increasing less quickly than output. Thus, between 1990 and 2004 there was a decline of 14% in the energy intensity of the service sector (as measured by final energy use per unit of value-added). Trends varied across countries, with both increases and decreases in energy intensity. The relative energy intensities of countries differ markedly, depending on whether value-added is calculated using market exchange rates or purchasing power parities.

▷   In most countries, total final energy use per unit of floor area fell less than energy use per unit of value-added. However, important differences were evident in the trends for electricity and fuels. For a group of ten IEA countries for which data are available, fuel use per unit of floor area fell by 12% whereas electricity rose by 17% per unit of floor area. As fuels are used mainly for space heating, this decline represents savings in space heating energy per unit of floor area.

▷   Without any reductions in energy intensity, final energy use in the service sector in the IEA19 would have been 17% or 2.8 EJ higher in 2004. $CO_2$ emissions would have been 280 Mt $CO_2$ higher. This represents an average annual reduction in energy intensity of 1.1% per year, which is substantially lower than the long-term trend of 2.3% per year between 1973 and 1990.

# PASSENGER TRANSPORT

## Scope

Passenger transport includes the movement of people by road, rail, sea and air. Road transport is further sub-divided into cars and buses. Only domestic air travel is included; international air travel is not covered.

## *Highlights*

Between 1990 and 2004, the overall energy efficiency of passenger transport in a group of 17 IEA countries improved by 0.5% per year. Without the energy savings resulting from these improvements, passenger transport energy consumption in the IEA17 would have been 7% higher in 2004 (Figure 6.1). This represents an annual energy saving in 2004 of 2.1 EJ, which is equivalent to 150 Mt of avoided $CO_2$ emissions.

Despite these savings, total final energy use in transport still increased by 25% between 1990 and 2004, driven primarily by rising car and air travel. The rate of energy efficiency improvement was much lower than in previous decades; an improvement of 1.0% per year was seen between 1973 and 1990.

**Figure 6.1 ▶** *Passenger Transport Energy Savings from Improvements in Energy Efficiency, IEA17*

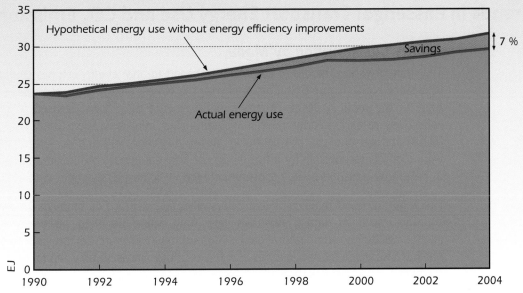

**6**

# Overview of Trends in Passenger Transport

Between 1990 and 2004, the volume of passenger travel (excluding international air travel) in 17 IEA countries[1], as measured by passenger-kilometres, increased by 31%. During the same period, total final energy use increased by 25% (Figure 6.2). This led to a decrease in the overall energy intensity of passenger transport. $CO_2$ emissions increased by 24%, reflecting a slight decrease in the carbon intensity of the energy mix.

**Figure 6.2** ▶ *Overview of Key Trends in Passenger Transport, IEA17*

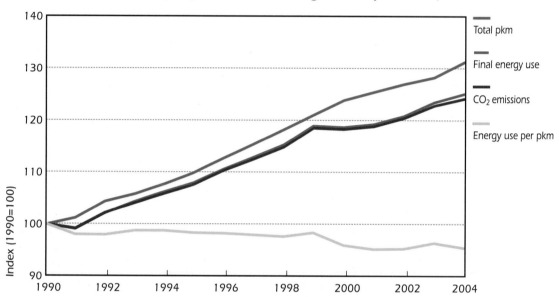

## Trends in Passenger Transport Energy Use and $CO_2$ Emissions

### Energy Consumption by Mode

The share of various modes of passenger transport in final energy use has changed little since 1990 (Figure 6.3). Cars are by far the largest energy user, responsible for 88% of the total in 2004. In this publication, the term "cars" is used to refer collectively to all light-duty vehicles including cars, mini-vans, sport utility vehicles (SUVs) and personal-use pick-up trucks. Buses, passenger rail and passenger ships were together responsible for a further 3% of final energy use. Approximately 10% of passenger transport energy consumption was in domestic passenger airplanes.

Passenger transport remains almost exclusively dependent on oil products. In a few countries, such as Japan, Australia and Italy, some liquefied petroleum gas (sometimes derived from natural gas) is used for road transport. Compressed natural gas has also made limited in-roads in a few countries, most notably Japan. Biofuels currently account for less than 1% of total passenger transport energy use, although their importance is growing in some IEA countries. Electricity for passenger rail travel is less than 1% of the total.

---

1. The 17 IEA countries included in the analysis of the passenger transport sector are Australia, Austria, Canada, Denmark, Finland, France, Germany, Greece, Ireland, Italy, Japan, the Netherlands, New Zealand, Norway, Sweden, the United Kingdom and the United States. These countries account for 90% of total IEA transport energy use.

**Figure 6.3** ▶ *Passenger Transport Energy Use by Mode, IEA17*

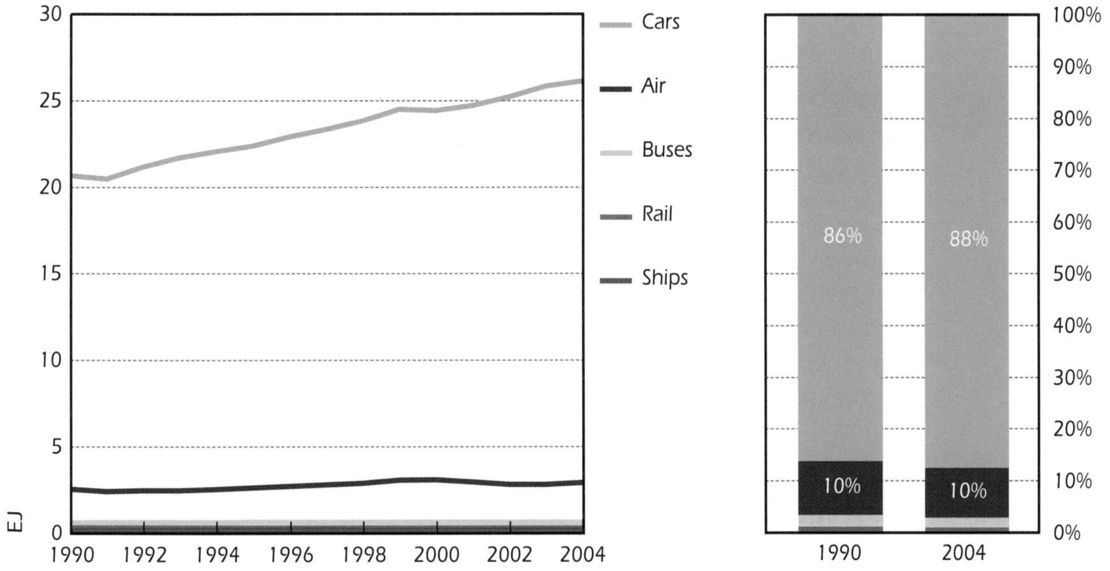

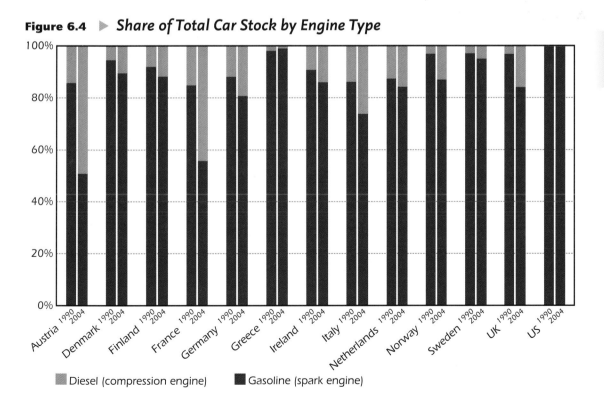

Although there has been little switching away from oil in passenger transport, the fuel mix has undergone some important changes in recent years. Most significant has been the "dieselisation" of cars in Europe, which started in the late 1980s and was boosted further by the widespread introduction of direct-injection diesel engines (Figure 6.4). The share of diesel vehicle sales in Europe more than doubled between 1997 and 2004; by 2004, nearly half of the cars sold in European IEA countries were equipped with diesel engines.

**Figure 6.4** ▶ *Share of Total Car Stock by Engine Type*

## CO₂ Emissions

Increasing energy use in passenger transport has led to a 13% rise in tailpipe $CO_2$ emissions per capita in the IEA17 since 1990. Increased emissions from cars were responsible for 97% of the total emissions growth in passenger transport. The strong link between energy use and emissions is due to the almost total reliance on oil-based fuels for cars, buses and airplanes. Together, these modes account for the large majority of passenger-kilometres travelled in IEA countries.

However, the growth in $CO_2$ emissions per capita has been uneven across the IEA (Figure 6.5). In some European countries, passenger transport $CO_2$ emissions per capita remained relatively stable or even decreased (e.g. in Finland, Germany and the United Kingdom), reflecting the rather small growth of travel per capita and improvements in the efficiency of vehicles.

**Figure 6.5 ▶ *Passenger Transport CO₂ Emissions per Capita***

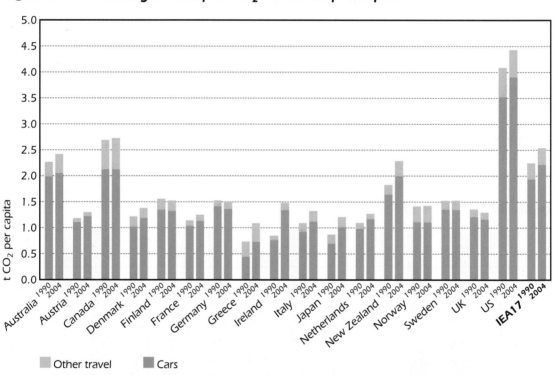

In contrast, Greece and Ireland had sharp increases in emissions per capita of 48% and 75%, respectively. This can be attributed to strong growth in passenger travel and, in particular, car use. Per capita $CO_2$ emissions from passenger transport also rose strongly in Japan and New Zealand. The significant increase in Japan was associated with higher fuel consumption in cars, coupled with a strong increase of car ownership and slightly rising travel per capita. Growing car ownership and, in particular, use were the main drivers for increased $CO_2$ emissions in New Zealand. In all countries, increasing domestic air travel accounted for most of the remaining rise in $CO_2$ emissions.

# Drivers of Energy Use in Passenger Transport

Passenger travel activity in the IEA increased steadily between 1990 and 2004, extending the trend that characterised the two previous decades. The strongest per capita increases were in car passenger travel (1.3% per year) and air travel (2.8% per year). In contrast, per capita passenger travel by bus and train decreased slightly over this period.

The level of passenger transport per capita varies widely from country to country (Figure 6.6). Those countries with a high density of population, such as Japan and the Netherlands, have significantly lower levels of travel per capita than low-density countries such as Australia, Canada and the United States. These latter countries have many fewer inhabitants per square kilometre and a much higher share of air travel. The United States has the highest levels of travel per capita (about 30 000 km per person per year), reflecting high rates of car ownership and utilisation.

**Figure 6.6** ▶ *Passenger Travel per Capita, All Modes*

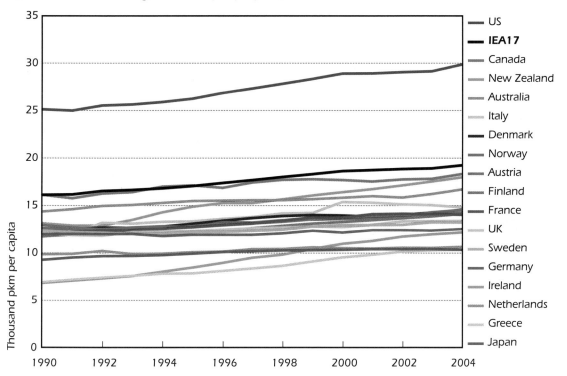

Trends in total passenger travel per capita also differ significantly across the IEA. Between 1990 and 2004, total passenger-kilometres per capita increased by just 5% in Sweden and 7% in Germany. The highest increases occurred in Greece (50%) and Ireland (79%). This reflects the fact that both Greece and Ireland had relatively low levels of passenger travel in 1990 and are now catching up with other IEA countries.

These changes in patterns of passenger travel are partly a consequence of growing wealth. As per capita incomes increase, people tend to migrate toward using faster, more flexible and more expensive modes of travel. For example, individuals "upgrade" from buses and trains (which have a decreasing share in many IEA countries) to cars or airplanes (the shares of which are increasing). With faster modes of transport, people also tend to travel further – *i.e.* they do not reduce the amount of time spent travelling. In reality, travel patterns are influenced by many diverse factors such as income, age profile, gender, household size, flexible working and leisure activities, as well as area-specific characteristics and local transport policies. Thus, it is important to keep in mind that individual travel patterns and modal shares can follow specific developments that do not match the aggregate picture.

Trends in the share of passenger transport by mode are shown in Figure 6.7. Cars clearly dominate the overall modal split in all IEA17 countries. On average, they accounted for 80% of total passenger-kilometres in both 1990 and 2004. Yet the share of car travel differs from country to country, reflecting diverse demographic and geographic characteristics, as well as different levels of provision for urban and intercity transport. Between 1990 and 2004, air travel was the fastest growing mode of passenger travel. Buses and trains each accounted for 5% of the total passenger travel in 2004, slightly lower than in 1990.

**Figure 6.7** ▶ *Share of Total Passenger Travel by Mode*

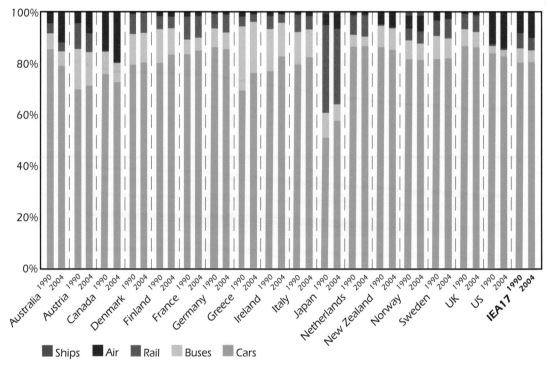

Japan stands out because of the large share of passenger-kilometres travelled by rail, which accounted for 34% in 1990 and 29% in 2004. This is due to a strong urban and regional rail system developed over the previous decades. The rail system in Japan was already quite mature well before 1990 and has not grown substantially in the past two decades. As other modes have become more popular in Japan, there has been a reduction in rail's share of passenger transport.

European countries are characterised by a significant share of public passenger transport (primarily buses and trains), especially if compared to Australia, Canada, and the United States. In these three countries, air passenger shares grew to well above 10% in 2004, making air travel second in importance to cars. However, it should be noted that the information for European countries excludes pan-European air travel (the data include only domestic passenger air trips).

Despite some variations in the share of passenger-kilometres among countries, cars are the main driver behind increased passenger travel across the IEA17. Once individuals and families own a car, two things tend to happen: they do more of their travel by car and their total travel rises significantly. With the exception of Canada, all countries saw increases in car ownership per capita between 1990 and 2004 (Figure 6.8). In 1990, most IEA countries had car ownership levels of between 0.30 to 0.45 cars per capita. Greece and Ireland, however, had significantly lower ownership rates at this time, of 0.17 and 0.23 cars per capita respectively. The only countries with ownership rates higher than 0.50 were New Zealand and the United States.

**Figure 6.8** ▶ *Car Ownership per Capita*

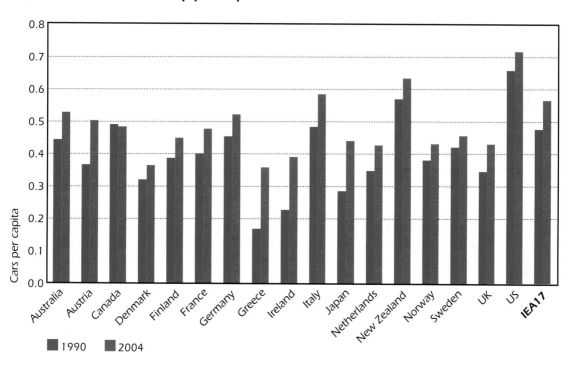

By 2004, no country within the IEA17 had an ownership level below 0.35 cars per capita. Greece and Ireland had the fastest growth in ownership, roughly doubling their 1990 levels. The increase for all other IEA countries was more modest, as car ownership levels were already relatively high in 1990. Most European countries are now in the range of 0.40 to 0.55 cars per capita. Italy is the exception, with an ownership of almost 0.60. This can be attributed to a historically strong development of the intercity road infrastructure and limited development of rapid urban transit rail systems, which creates higher reliance on cars.

Economic growth has a strong influence on car ownership. Rising incomes are almost always associated with increases in car ownership, although there are significant variations in this relationship from country to country (Figure 6.9). This trend helps to explain why the United States, which had the highest levels of personal consumption expenditure, also had the highest car ownership level (above 0.70 in 2004). However, the rate of growth in car ownership tends to slow down once average expenditure rises

**Figure 6.9** ▶  *Car Ownership per Capita and Personal Consumption Expenditures,*
*1990 – 2004*

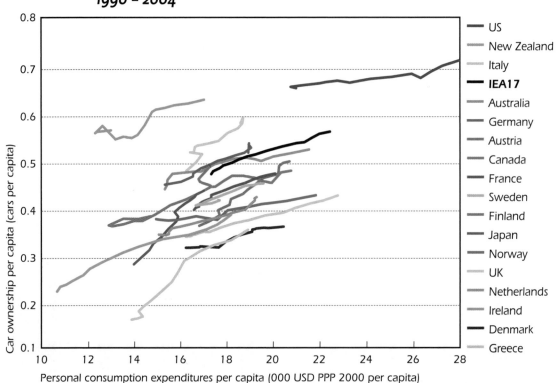

above a certain level. Indeed, the rise in car ownership in United States since 1990 has been among the lowest in the IEA. This suggests that a saturation point might eventually be reached, perhaps somewhere around 0.7 – 0.8 cars per capita.

Car travel per capita is also closely related to incomes (Figure 6.10). All IEA17 countries have experienced a steady increase in car travel per capita as personal consumption expenditures have increased. This stems from the tendency noted earlier that as people grow richer, they travel further and use faster, more flexible and more expensive modes of transport.

However, there are significant variations among IEA countries in the relationship between car travel per capita and personal expenditures. Some countries, such as New Zealand and the United States, have much higher than average levels of travel by car because of high ownership levels and relatively low densities of population, leading to longer average journeys. Japan has historically had the lowest ratio of car travel, reflecting its high population density, extensive public transport infrastructure and below average levels of car ownership. The patterns in Europe reflect a combination of factors such as population density, fuel tax increases and other policy initiatives to curb the growth of car travel.

Fuel prices for passenger transport are an important factor influencing travel demand, mode choice and energy intensity (Figure 6.11). Fuel prices vary considerably across IEA countries, with gasoline prices in 2004 ranging from USD 0.45 to USD 1.26 per litre (using 2000 PPP exchange rates). Variations in fuel taxes are the main reason for this wide gap. Three groups of countries can be identified: the United States, with the lowest prices (and the lowest tax levels);

**Figure 6.10** ▶ *Car-kilometres per Capita and Personal Consumption Expenditures,*
*1990 – 2004*

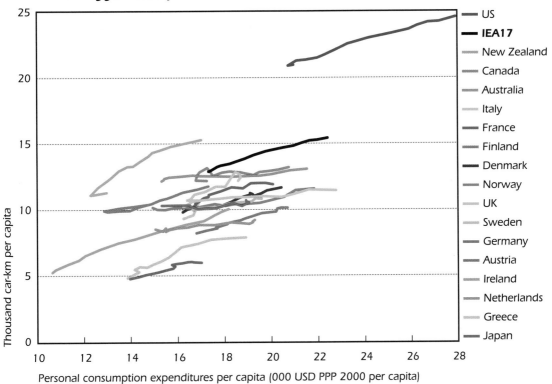

Personal consumption expenditures per capita (000 USD PPP 2000 per capita)

**Figure 6.11** ▶ *Trends in Retail Gasoline Prices in Real Terms*

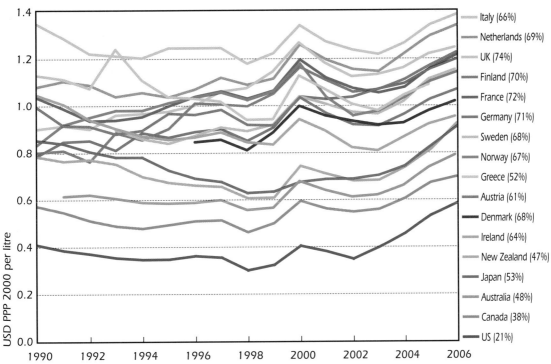

Note: Figures in parentheses refer to the percentage of taxes in the total retail gasoline price for non-commercial use in 2004.
Source: IEA/OECD *Energy Prices & Taxes*.

European countries, with generally high prices (and high taxes); and, in the middle, other IEA countries such as Australia, Canada, Japan and New Zealand.

The evolution of oil prices is the key element underlying the variation in fuel prices over time. Because they have not substantially changed their fuel taxes since 1990, price trends for Japan and the United States most strongly reflect this underlying development. In many other countries (and notably in Europe), the oil price influence is evident over time even though price variations are cushioned somewhat by the level of fuel taxes. Following a peak in the early 1980s, oil prices declined in real terms for the next 15 to 20 years (see Figure 2.12 in Chapter 2). As a result, gasoline prices in most IEA countries were either flat or declining slightly during much of the 1990s. After 1998, oil prices rose again, in turn, pushing up gasoline prices with a particularly sharp increase in 1999 and 2000. Since this time, gasoline prices in most countries have remained higher in real terms than in the 1990s.

The impact of price changes on fuel costs per vehicle-kilometre can be calculated by combining fuel prices with the average fuel intensities of the car fleet (see next section). Fuel costs per vehicle-kilometre provide an indication of the "cost of driving", but they do not take into account the vehicle purchase price. Between 1990 and 2004, the cost of driving showed a somewhat similar pattern to gasoline prices – *i.e.* a downward trend for much of the 1990s, followed by a temporary peak in 2000 and then a sustained increase after 2003 (Figure 6.12). However, in many countries, the increases in fuel costs per vehicle-kilometre since 1999 have been less severe. This is particularly true in Europe, where high fuel and vehicle taxes, pollutant emission regulations and an increased share of diesel cars all contributed to a more efficient car fleet. Together, these factors helped offset some of the rise in fuel prices.

**Figure 6.12** ▶ *Fuel Costs per Vehicle-kilometre for Cars*

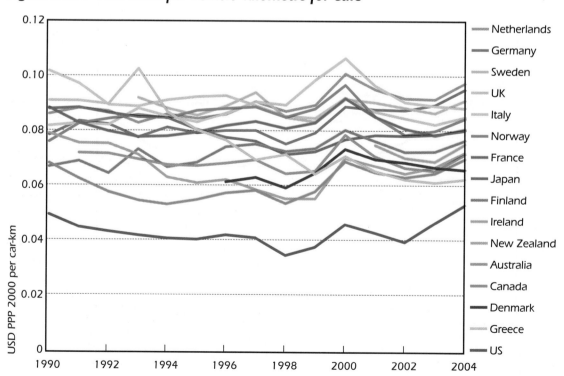

Apart from the United States, the variation among countries in fuel costs per vehicle-kilometre in 2004 is lower than for fuel prices, lying in a range between USD 0.06 and USD 0.10 per kilometre driven. This convergence of fuel costs per kilometre seems to reflect a degree of interdependence between cost and efficiency. Over time, drivers respond to higher fuel prices by buying more efficient vehicles and, thus, bring their overall driving costs into an acceptable range. Even in the United States, the trend since 1998 has seen driving costs move much closer to other IEA countries.

The relationship between fuel price and per capita fuel use effectively separates IEA countries into distinct groups (Figure 6.13). The United States has almost twice the car fuel use per capita compared to the next group of countries, which includes Australia, Canada and New Zealand. The European countries are characterised by high fuel prices (two to three times higher than the United States) and a fuel use per capita that is often less than one-third of the United States. Japan also has very low fuel consumption per capita, even though fuel prices are about one-third less than in many European countries. Australia, Canada and New Zealand fall between Japan and the United States in terms of both fuel price and fuel use per capita.

**Figure 6.13** ▶ *Fuel Use per Capita versus Fuel Prices, 2004*

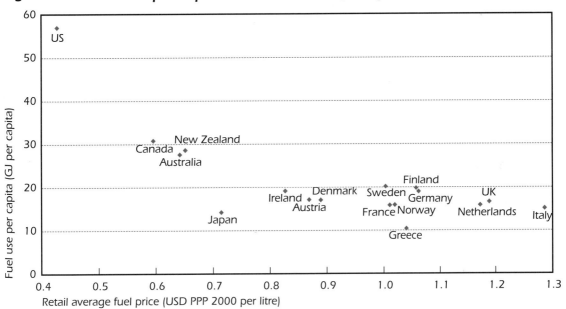

Looking at Europe and Japan, there is no clear relationship between fuel prices and fuel use per capita. Other factors explain most of the variation including country size, income, vehicle prices, taxes and ownership levels, as well as the availability and extent of mass transit. For example, despite a lower average fuel price than countries in Europe, Japan has the second lowest energy use per capita. This can be attributed to the high availability and extent of mass transit, and to low travel per capita (Japan is densely populated and travel distances are shorter than in many other countries).

Australia, Canada and the United States are geographically the largest countries and are much less densely populated than Japan and Europe. These factors partially explain why their energy use per capita is higher. However, size does not explain the differences between Australia, Canada and the United States. This gap is influenced by the significantly lower energy prices and higher incomes in the United States.

Splitting car fuel use per capita into car travel per capita and car fuel intensity (measured as gasoline equivalent litres per 100 km) provides some explanation for the relationship between fuel use and price. The top part of Figure 6.14 indicates only a weak correlation between low fuel prices and high car travel rates (Australia, Canada and the United States). European countries have relatively similar levels of car travel per capita, despite significant variations in real fuel prices. Japan has the lowest travel rates of all countries, even though the fuel price is relatively low. This highlights how car fuel use is influenced by Japan's relatively small geographical size and well-developed mass transit system.

**Figure 6.14** ▶ *Passenger Travel per Capita and Car Fuel Intensity versus Average Fuel Price, 2004*

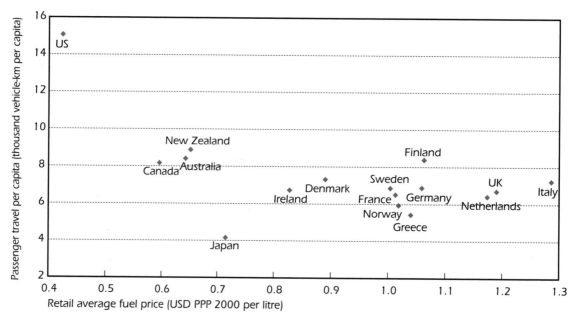

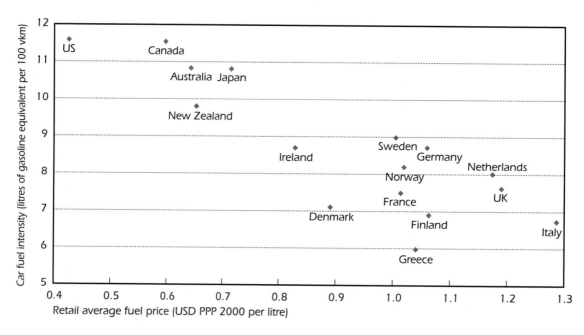

In contrast, the lower part of Figure 6.14 shows a relatively strong correlation between higher fuel prices and lower car fuel intensity. Interestingly, in this case Japan no longer falls outside of the general trend: its relatively high fuel intensity and low fuel price place it within the same range as Australia and Canada. Thus, Japan's low car fuel use per capita relative to fuel price results from modest car use, not from low fuel intensity. Australia and Canada, which represent mid-range countries for per capita car fuel use, are close to the United States in fuel intensity but more aligned to European countries in terms of car travel per capita.

# Changes in Aggregate Energy Intensity

The final energy intensity of most passenger transport modes, as measured by energy use per passenger-kilometre, declined between 1990 and 2004 (Figure 6.15). Air travel intensity showed the largest reduction of 28%, which can be attributed to a combination of several factors. The energy efficiency of aircraft engines has been improving, thanks to significant advances achieved in computational fluid dynamics, advanced materials and the development of software tools for engine design. Increased use of light materials (*e.g.* composites) resulted in lower lift requirements and the design of larger aircraft that demand less energy while transporting more passengers. At the same time, increases in average load factors effectively lowered energy intensities at the margin: filling an empty seat moves one more passenger a long distance with almost no change in energy use.

Energy intensities for other modes of passenger transport have not declined as much, even though significant technological potential is available. Average car energy intensity decreased by 4% across the IEA17. Several factors influenced this development: the trends in the efficiency of new cars; the (generally lower) efficiency of the vehicles that were scrapped; and the evolution of passenger load factors (the lower the load factor, the higher the energy intensity per passenger-kilometre). Until

**Figure 6.15** ▶ *Energy per Passenger-kilometre by Mode, IEA17*

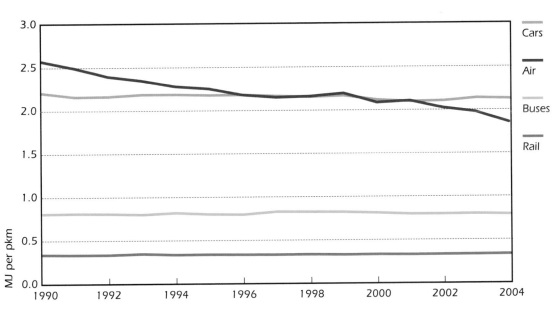

the late 1990s, there was little improvement in new car energy intensities (indeed, in some countries they actually increased) and load factors declined only slightly. This combination of factors led to a modest fall in the average energy intensity of the car fleet during the early and mid-1990s. In subsequent years, increased diffusion of more efficient vehicles – particularly in Europe and Japan – led to a faster rate of reduction in the energy intensity of passenger transport.

Rail and bus travel have always been relatively efficient in comparison with other transport modes. However, their energy intensity has not declined much in recent years for several reasons. For example, significant improvements in the efficiency of diesel engines for buses were offset by declining load factors. In the rail sector, the increasing diffusion of high-speed rail connections in Europe increased energy demand per vehicle (in comparison to conventional trains), although this was counterbalanced by higher occupancy rates.

Figure 6.16 shows the trends in energy use per passenger-kilometre by country, aggregated across all modes. These trends are influenced by both the energy intensity of each mode and by the share of that mode in a particular country. For most countries, energy use per passenger-kilometre is declining. Reductions in the energy intensity of individual modes (Figure 6.15) have been more than enough to offset the impact of increasing shares of car and air travel (Figure 6.7), which are more energy intensive. The only exceptions are Japan and the Netherlands, where energy per passenger-kilometre has increased. Japan had one of the lowest energy intensities for passenger transport in 1990, but has since shown a significant increase. This can be attributed to both a falling share of rail (at the expense of cars) and to an increase in the energy intensity of cars (at least until recent years). For the Netherlands, higher levels of car ownership, coupled with virtually no change in their energy intensity, were the main reasons for a rise in energy per passenger-kilometre.

**Figure 6.16** ▶ *Energy per Passenger-kilometre Aggregated for All Modes*

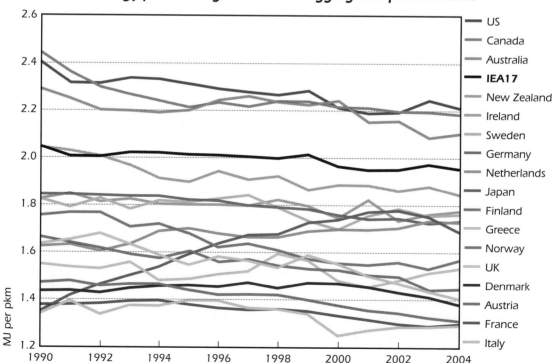

Some convergence among countries is evident in the level of energy use per passenger-kilometre between 1990 and 2004, yet very significant differences remain. This can partly be explained by differences in the modal splits from country to country. However, most of the variation reflects different levels of energy intensities for individual modes, particularly for cars.

A more detailed analysis of cars reveals wide variation in the levels and trends of average "on-road" fuel intensities (Figure 6.17). The results reflect a number of unrelated factors such as vehicle technologies and the effect of driving conditions. Increased urban traffic congestion, aggressive driving, higher speeds on highways and more in-car amenities (such as air conditioning) have all contributed to widening the gap between fuel economy ratings obtained from vehicle tests and actual on-road performance.

**Figure 6.17** ▶ *Average Fuel Intensity of the Car Stock*

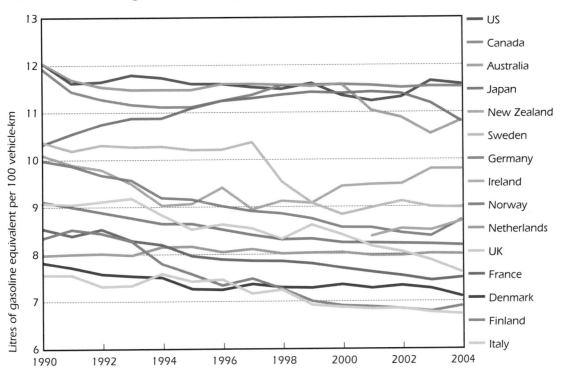

The average fuel intensities of cars decreased in most countries between 1990 and 2004. In Europe, this was due to a combination of factors. The 1990s were characterised by the widespread diffusion of vehicles equipped with electronic control systems for fuel management and by stronger consumer demand for more efficient cars – a reaction to high fuel prices. Since the early 2000s, intensities declined further in Europe as a result of increased sales of direct-injection diesel cars (which are about 20 – 25% more efficient than their gasoline equivalents).

Despite a small decrease, the fuel intensity of cars in North America remained higher than in other IEA countries, at nearly 11.5 gasoline equivalent litres (litres of diesel and alternative fuels have been converted to their gasoline equivalent based on their relative energy content per litre). High levels of fuel intensity also characterised Australia and Japan. In Japan, until the late 1990s, efficiency improvements of new vehicles had been offset by congestion-related effects.

## Box 6.1

# Trends in the Weight of New Passenger Cars

The weight of new passenger cars has increased in all IEA countries over the past 10 years (Figure 6.18). This growth was particularly strong in the United States, where weight increases were driven by a growing share of light trucks. New cars have also become progressively heavier in Europe for three reasons: manufacturers are now offering new models that are larger than their predecessors; large models (such as SUVs) make up a higher share of new car sales; and the significant switch to diesel vehicles, which are generally heavier than gasoline cars offering the same performance.

In contrast to the United States and Europe, Japan reported a much lower rate of increase in the weight of new cars, mainly because of the effective regulation of fuel efficiency under the Top Runner programme. In addition, between 1990 and 2005, Japan saw an increasing share of very small mini-cars (close to 30% of the total car sales in 2005), which has helped offset the effect of an increased demand for large vehicles.

**Figure 6.18 ▶** *Average Weight of New Passenger Cars*

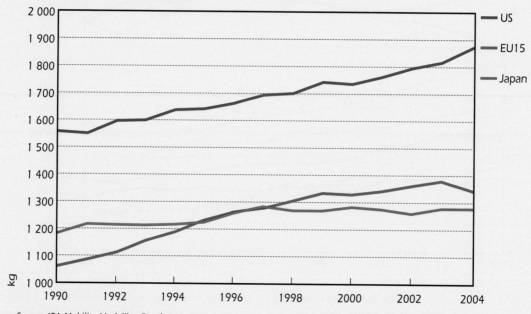

Source: IEA Mobility Modelling Database.

Note: EU15 is a group of 15 European Union countries

Evolving safety features have also played a role in the increase in vehicle weight, although the association between heavier vehicles and improved safety is not straightforward. The safety characteristics of a vehicle depend on its structural performance, not just on how heavy it is. A number of light-weight materials allow car manufacturers to improve overall structural performance of vehicles while maintaining, or even reducing, their weight. Moreover, reducing the size (and thereby lowering weight) of larger vehicles would improve overall passenger safety in accidents involving both large and small vehicles.

Examining the data on the test results for fuel intensities of new cars provides a better understanding of the role of new technologies in reducing on-road fuel intensities (Figure 6.19). In the early 1990s in many European countries, the fuel intensity of new cars remained flat or increased. However, significant reductions have been achieved since the mid-1990s. To a large degree, the change in trends coincided with an increase in the market share of diesel cars in Europe (see Figure 6.4). The subsequent introduction of direct injection technology made diesel cars significantly more efficient than gasoline cars, even taking account of the weight penalty (see Box 6.1). The voluntary agreement between car manufacturers and the European Commission also played an important role in achieving lower energy intensities for new cars.[2]

**Figure 6.19 ▶ *Trends in New Car Fuel Intensity***

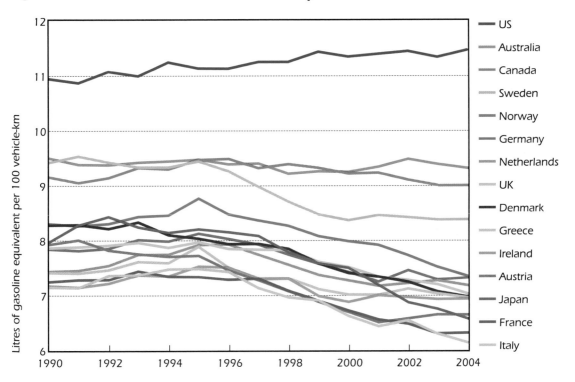

Source: IEA Mobility Modelling Database.

The trend for new car fuel intensities in Japan was similar to that of Europe until the late 1990s, when the government introduced the Top Runner programme. This scheme encourages manufacturers to improve the efficiencies of all vehicles to the level of the best performer in each market segment. The programme proved particularly effective for light duty vehicles. As a result, improved technologies have become more widespread such as variable valve timing and, more recently, continuously variable transmission. The number of hybrid vehicles has also increased significantly; by 2005, they represented more than 4% of total new car sales.

---

2. The European Commission signed a voluntary agreement on fuel economy improvements with car manufacturers in Europe, Japan and Korea. The aim is to achieve a 25% reduction (from 1995 levels) in $CO_2$ emissions from new passenger cars in the European Union (to an average of 140 g/km) by 2008/9.

Unlike in Europe and Japan, which have both experienced declines, the evolution of trends for new car fuel intensity has been slowly upwards in the United States. This was due to the increasing sales of personal light trucks, which are much heavier than ordinary passenger cars and have significantly higher energy intensities. The United States has a much higher share of more fuel-intensive personal light trucks than any other country; as a result, the average new (and fleet) fuel intensity for all cars is the highest of any IEA country. Also in contrast to Europe and Japan, there was little policy pressure to reduce fuel intensities: since 1990, there has been no change to the Corporate Average Fuel Economy (CAFE) standards for passenger cars.[3]

## Impact of Changes in Structure and Intensity

Figure 6.20 decomposes the trends in energy use for passenger transport across all modes between 1990 and 2004 into changes due to activity, modal structure and modal intensities (as described in Annex A).

**Figure 6.20** ▶ *Factors Affecting Passenger Transport Energy Use, IEA17*

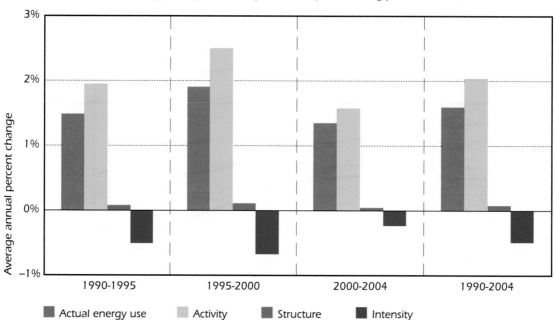

Energy use across all modes increased substantially between 1990 and 2004. Not surprisingly, increased activity (*i.e.* more passenger-kilometres) was the most important factor driving up energy use. Structural changes – particularly the increase in the shares of cars and air transport – also tended to increase energy use, but their impact was much more limited. These upward pressures on energy use were only

---

3. The Corporate Average Fuel Economy (CAFE) is the required average fuel economy for a vehicle manufacturer's entire fleet of passenger cars and light trucks manufactured for sale in the United States, for each model year. There are separate average fuel economy standards for passenger cars and light trucks (light trucks include those with a gross vehicle weight rating of 8 500 pounds [4 000 kg] or less).

partially offset by reductions in energy intensity, which fell at an average rate of 0.5% per year. The increasing share of diesel cars, particularly in Europe, was an important reason behind the overall fall in energy intensity (diesel engines are more efficient than those using gasoline). However, the rate of energy intensity reduction since 1990 was much lower than in previous decades; an average fall of 1.0% per year was seen between 1973 and 1990 (see Annex C).

Growth in passenger travel was strongest between 1995 and 2000, as a result of lower fuel prices in many IEA countries and a strong global economy. Energy intensity reductions were more significant between 1990 and 2000 than between 2000 and 2004, even though it might have been expected that higher oil prices and the introduction of more efficient vehicles (particularly cars) would have accelerated reductions in energy intensity in the latter time period. However, it should be noted that the energy intensity effect is the result of a complicated interaction among vehicle efficiencies, operating conditions and load factors. Thus, the combined effect is often hard to predict.

Trends in $CO_2$ emissions from passenger transport closely follow the changes in energy consumption (Figure 6.21). This is mainly due to transport's heavy dependence on oil products, which are relatively similar in terms of $CO_2$ intensities. However, some differences exist between the tailpipe $CO_2$ emission characteristics of diesel and gasoline vehicles. A higher diesel share is reflected in an increased fuel mix effect (as a result of higher $CO_2$ emissions per unit of energy than gasoline), which tends to increase $CO_2$ emissions across all time periods. The blending of non-oil based alternative fuels with either gasoline or diesel is reflected in a change in the carbon intensity effect. This change was negative between 1990 and 2004, reflecting an increased use of biofuels, which produce no net $CO_2$ emissions from the tailpipe.

**Figure 6.21** ▶ *Decomposition of Changes in Passenger Transport $CO_2$ Emissions, IEA17*

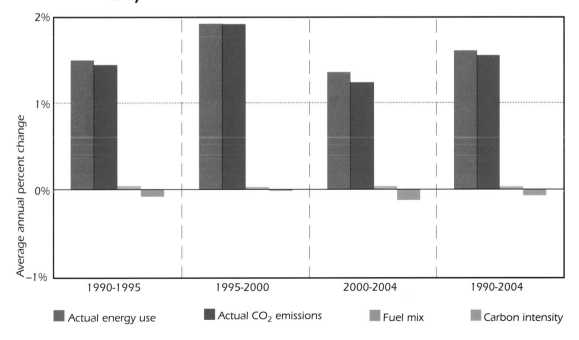

Several factors influence car energy use across IEA countries (Figure 6.22). All countries, with the exception of Canada, showed increases in car ownership. Greece and Japan showed the strongest growth, albeit rising from comparatively low ownership levels in 1990. For most countries, the growth in car ownership tended to increase per capita car energy consumption by about 1% per year. The impact of car usage (*i.e.* the distance driven by each car) on per capita energy consumption is more varied across countries. In most countries, the distance travelled by each car increased. However, car usage actually fell in six countries: Australia, Germany, Japan, the Netherlands, Norway and the United Kingdom. In these countries, the trend toward households owning more than one car means that journeys are shared between cars. As a result, travel per car tends to fall.

Together, car ownership and usage give the total distance travelled per capita. For most countries, reductions in the fuel intensity of cars were not sufficient to offset the increases in car ownership and car use. Thus, car energy use per capita increased in many IEA countries. The exceptions to this were Finland, Germany, Norway and the United Kingdom. In these countries, the effect of significant reductions in energy intensity were augmented by falling car usage (except in Finland, which showed a small increase), which more than offset increases in car ownership. In Japan, fuel intensities increased slightly as the impacts of increased congestion offset improvements in engine efficiency.

**Figure 6.22** ▶ *Decomposition of Changes in Car Energy Use per Capita, 1990 – 2004*

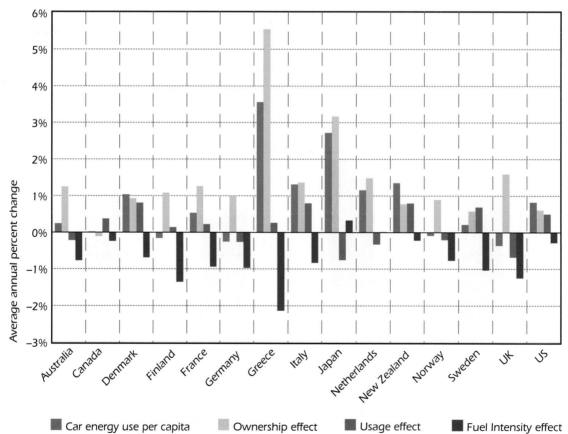

Note: Austria and Ireland are excluded due to the lack of complete time series data for vehicle-kilometres.

# Energy and CO$_2$ Savings

Figure 6.23 shows the energy savings resulting from a decline in energy intensity in passenger transport. By 2004, annual energy savings totalled 2.1 EJ or 7% of passenger transport final energy use in that year. The impact of energy intensity improvements on CO$_2$ emissions were also significant, with a saving in 2004 of 150 Mt CO$_2$ (7% of actual CO$_2$ emissions in 2004).

**Figure 6.23** ▶ *Passenger Transport Energy and CO$_2$ Emissions Savings from Reductions in Energy Intensity, IEA17*

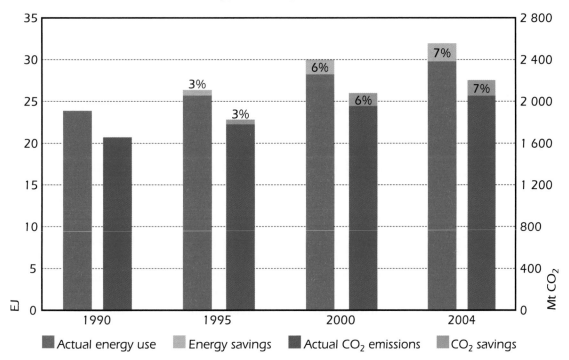

# Summary

Examination of the passenger transport sector of 17 IEA countries over the period 1990 to 2004 reveals the following key findings:

▷ Total final energy consumption in domestic passenger transport (excluding international air travel) increased by 25%; CO$_2$ emissions rose 24%. These trends were driven largely by a 31% increase in passenger travel, measured by the number of passenger-kilometres.

▷ Cars are responsible for the largest share of energy use in domestic passenger transport, accounting for 88% of the total in 2004. In most countries, cars remain almost exclusively dependent on oil. Europe has seen a significant increase in the penetration of diesel as a fuel. The share of biofuels is still small, but is also growing in some countries.

▷ The growth in passenger travel is related to higher incomes and is mainly associated with the use of cars and airplanes. Car ownership continues to increase, particularly in those countries that had relatively low ownership levels in 1990. Domestic air travel is the fastest growing mode of transport, and was 61% higher in 2004 than in 1990. In contrast, the shares of bus and rail travel have declined.

▷ Levels of passenger travel per capita vary widely among countries. Those countries having a high population density, such as Japan and the Netherlands, show significantly lower levels of passenger travel per capita than larger, less dense countries, such as Australia, Canada and the United States. The modal shares of passenger transport also vary significantly from country to country. Japan has a high share of rail transport. Many European countries have a significant share of public transport, particularly in comparison to Australia, Canada, New Zealand and the United States.

▷ Fuel prices for passenger transport are an important driver of mode choice and fuel intensity. After-tax fuel prices vary by a factor of almost three across IEA countries, with the difference in fuel taxes being the main reason. In most countries, gasoline prices were relatively flat or declining until the late 1990s. Since 1998, they have increased in response to higher oil prices.

▷ The energy intensities of all passenger modes have declined, with a particularly strong decrease for domestic air travel. Overall, the energy intensity of passenger transport reduced at an average rate of 0.5% per year. However, this rate of reduction was much lower than in previous decades; a fall of 1.0% per year was seen between 1973 and 1990.

▷ Across the IEA, car energy intensity declined by only 4% on average. However, there are strong regional differences. From the mid-1990s, there have been significant reductions in the fuel intensity of new cars in Europe, partly as the result of a voluntary agreement between car manufacturers and the European Commission. The trend in Japan has been similar, but with a turning point in the late 1990s corresponding to the introduction of the Top Runner programme. In the United States, the evolution of fuel economy trends has been slowly upwards.

▷ Without the energy savings resulting from reductions in energy intensity, final energy consumption in passenger transport would have been 7% higher in 2004. This represents an annual energy saving of 2.1 EJ and 150 Mt of avoided $CO_2$ emissions.

# FREIGHT TRANSPORT

## Scope

Freight transport covers the domestic haulage of goods by trucks, rail, and ships and barges. It excludes air freight transport because of a lack of data separating domestic and international journeys. Pipelines are also excluded.

## *Highlights*

Between 1990 and 2004, the overall energy efficiency of freight transport in a group of 17 IEA countries improved by 0.6% per year. Without the energy savings resulting from these improvements, freight transport energy consumption in the IEA17 would have been 9% higher in 2004 (Figure 7.1). This represents an annual energy saving of 1.1 EJ in 2004, which is equivalent to 80 Mt of avoided $CO_2$ emissions.

Despite these savings, total final energy use in freight transport still increased by 25% between 1990 and 2004, driven largely by increased trucking. The rate of improvement in energy efficiency is slightly lower than in previous decades; it averaged 0.7% per year between 1973 and 1990.

**Figure 7.1** ▶ *Freight Transport Energy Savings from Improvements in Energy Efficiency, IEA17*

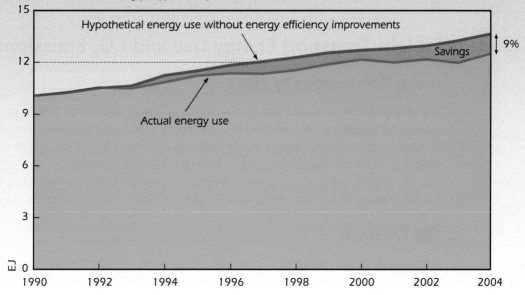

# Overview of Trends in Freight Transport

Between 1990 and 2004, the total volume of freight haulage (excluding air transport), as measured by tonne-kilometres, increased by 32% for 17 IEA countries.[1] The increase in total final energy use was less, at 25% (Figure 7.2). Thus, the overall final energy intensity of freight transport decreased. At the same time, $CO_2$ emissions increased by 24%, reflecting a slight decrease in the carbon intensity of the fuel mix.

**Figure 7.2** ▶ *Overview of Key Trends in Freight Transport, IEA17*

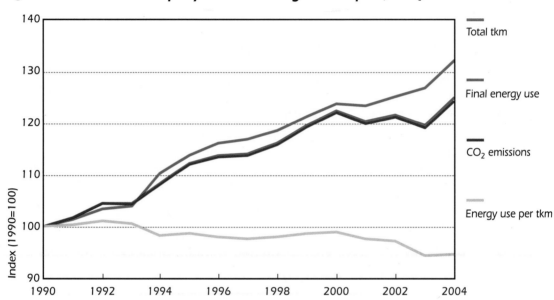

# Trends in Freight Transport Energy Use and $CO_2$ Emissions

## Energy Consumption by Mode

The strong growth in freight energy use was due almost entirely to higher energy demand for trucking, which increased by 33% (Figure 7.3). Trucks increased their share of total freight transport energy consumption to 82% in 2004. Total final energy consumption for rail freight increased by 16%, but its share of energy use declined to 6%. In contrast, energy use for water freight declined by 12%. Oil dominates the freight transport sector, accounting for 99% of total final energy consumption.

## CO2 Emissions

The pattern of $CO_2$ emissions from freight transport reflects the dominance of trucking. Figure 7.4 presents $CO_2$ emissions from freight haulage per unit of GDP for the IEA17 countries, split into truck and other (rail and shipping). There is considerable variation among countries, which reflects a combination of three factors: the volume of freight haulage per GDP; the share of the various freight

---

1. The 17 IEA countries included in the analysis of the freight transport sector are Australia, Austria, Canada, Denmark, Finland, France, Germany, Greece, Ireland, Italy, Japan, the Netherlands, New Zealand, Norway, Sweden, the United Kingdom and the United States. These countries account for 90% of total IEA transport energy use.

**Figure 7.3 ▶** *Freight Transport Energy Use by Mode, IEA17*

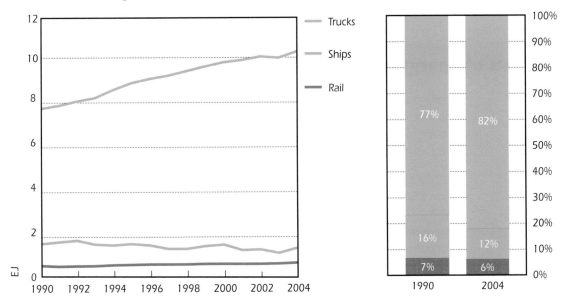

modes; and the energy intensity (energy per tonne-kilometre) of each mode. Canada has the highest emissions per GDP, largely as a result of long haulage distances. In contrast, Austria and Sweden have much lower emission intensities due to a combination of significantly shorter haulage distances and lower than average energy intensities. In 2004, rail and shipping accounted for a significant portion of

**Figure 7.4 ▶** *Freight CO₂ Emissions per GDP, IEA17*

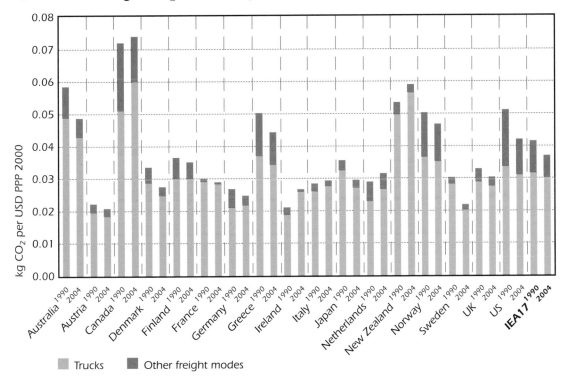

7

$CO_2$ emissions in the United States (27%), Norway (25%) and Greece (23%). All IEA17 countries, except Norway, have experienced an increase in the share of emissions from trucks between 1990 and 2004.[2]

# Drivers of Energy Use in Freight Transport

Freight transport energy use is driven by the growth in total freight haulage. Economic activity is associated with the movement of raw materials, intermediary products and final consumer goods. Thus, there is a strong correlation between increases in freight transport and GDP growth.

The relationship between total freight tonne-kilometres and GDP is illustrated in Figure 7.5. Between 1990 and 2004, total freight haulage increased by 32% in the IEA17 – *i.e.* less than GDP (which increased by 40% in the IEA17). Overall, freight tonne-kilometres per unit of GDP declined by 0.4% per year, largely due to reductions in the freight intensity of Japan, the United Kingdom and the United States. Conversely, many of the IEA17 countries experienced an increase in the freight intensity of their economies. Ireland, in particular, experienced rapid growth

**Figure 7.5** ▶ *Total Freight Tonne-kilometres per GDP, All Modes*

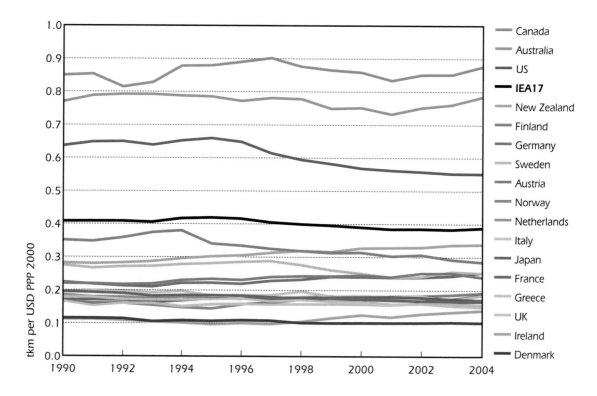

2. Figure B.6 in Annex B shows the same indicator as in Figure 7.4, re-calculated with GDP converted to USD using market exchanges rates (MER). This approach does not change the trends seen for each country (See Chapter 2, Box 2.2). However, when calculated using GDP at MER, New Zealand has a slightly higher energy intensity than Canada and Japan has a somewhat lower energy intensity than either Austria or Sweden.

of 1.6% per year between 1990 and 2004, albeit starting from a low level. Australia, Austria, Canada, Germany, New Zealand and Norway also experienced increases in their freight intensities, ranging from 0.1% to 1.3% per year.[3]

The relationship between freight transport by trucks and GDP is particularly strong (Figure 7.6). In contrast to the decline in total freight intensity of GDP, overall trucking intensity in the IEA17 increased by 0.5% per year between 1990 and 2004. Canada has seen a particularly rapid increase in trucking per unit of GDP over this period, of 3.1% per year. Austria, Australia, Germany, Ireland, New Zealand and Norway also experienced significant increases in trucking intensity. In contrast, a number of countries saw declines over the same period including Denmark, Finland, Greece, Italy, Sweden and the United Kingdom.

**Figure 7.6** ▶ *Truck Freight Tonne-kilometres per Capita versus GDP per Capita, 1990 – 2004*

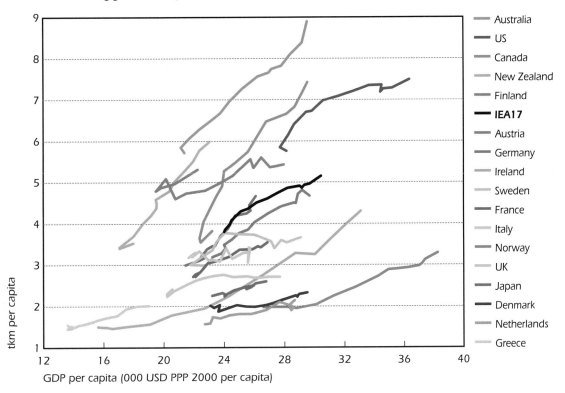

Calculating freight tonne-kilometres per capita is an alternative way of examining the level of freight transport in a given country. Between 1990 and 2004, this measure increased in all of the IEA17 countries (Figure 7.7). Levels of freight haulage per capita in Australia, Canada and the United States are significantly higher than in Europe, Japan and New Zealand. This reflects the large physical size of the first three countries, as well as their significant production and long-distance transport of raw materials.

---

3. Figure B.7 in Annex B shows the same indicator as in Figure 7.5, re-calculated with GDP converted to USD using market exchanges rates (MER). While this does not alter the trends in freight energy intensity, the relative positions of some countries do change. Using GDP expressed at MER, Canada, Australia and the United States still have the highest freight tonne-kilometres per GDP and the freight intensity of Denmark remains the lowest.

**Figure 7.7 ▶** *Total Freight Tonne-kilometres per Capita by Mode and Country*

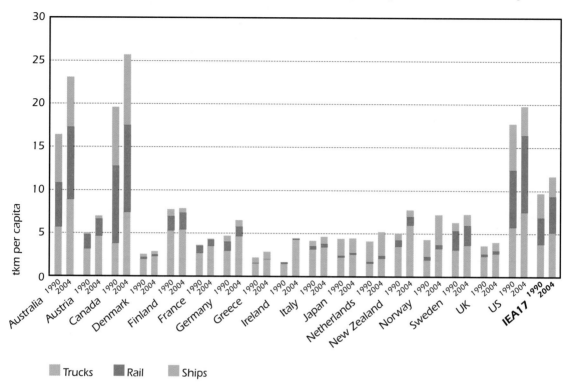

Australia and Canada experienced the largest absolute increases in total freight tonne-kilometres per capita between 1990 and 2004. Growth rates in freight haulage per capita have been highest in Ireland, Norway and New Zealand, although it should be noted that these countries started from relatively low levels in 1990.

Truck tonne-kilometres per capita increased in all IEA17 countries, albeit only modestly in Finland and Italy. Truck freight haulage was the most rapidly growing mode in Canada, Denmark, France, Germany, Ireland, Japan and New Zealand. In Ireland, trucking per capita almost tripled; in Canada, it nearly doubled. In 2004, Ireland had the highest percentage of tonne-kilometres transported by truck, at 96%. The share of trucking is also high in many other European countries, as well as in New Zealand.

Rail freight haulage per capita increased more quickly than other modes in five countries: Australia, Finland, the Netherlands, the United Kingdom and the United States. Rail carries the largest share of freight in the United States, with 45% of the total in 2004. Although the share of rail has fallen in Canada, from 46% to 39%, it is still high compared to the IEA average. Australia experienced the most rapid growth in freight tonne-kilometres transported by rail, with a 4.8% expansion in tonne-kilometres per year between 1990 and 2004. In contrast, a number of countries – Denmark, France, Greece, Ireland and Japan, in particular – experienced a decline in the absolute value of tonne-kilometres transported by rail.

Shipping tonne-kilometres per capita increased the fastest of the three modes in Austria, Greece, Italy, Norway and Sweden. Shipping is now the largest single freight haulage mode in Norway and the Netherlands. It is also important for domestic freight movement in Australia, Canada, Greece, Japan and the United Kingdom.

Fuel prices – and automotive diesel prices in particular – are a key element of freight transport costs. Thus, they may be expected to have an impact on levels of freight haulage and energy use. Automotive diesel prices vary widely among the IEA17 countries (Figure 7.8). Prices in Europe are significantly higher than in those in other IEA countries, largely due to higher taxes. Apart from taxes, diesel prices are most strongly influenced by the evolution of crude oil prices over time, as can be seen by analysing how diesel prices have corresponded with an upward trend in oil prices since 1998 (see Chapter 2 for a more detailed discussion on oil prices).

**Figure 7.8** ▶ *Trends in Automotive Commercial Diesel Prices in Real Terms*

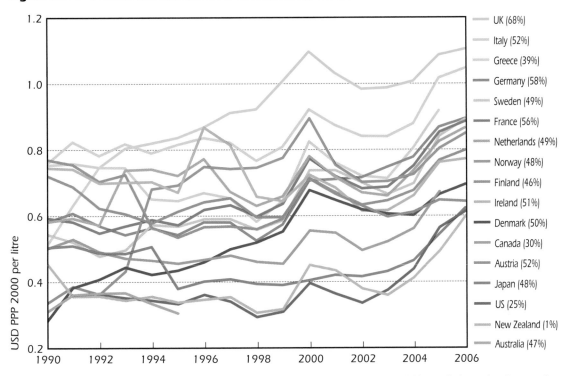

Notes: Excludes refundable value-added tax. Figures in parentheses refer to the percentage of non-refundable taxes in the total retail automotive diesel price for commercial use in 2004. Refundable taxes are excluded from the total retail price.

Source: IEA/OECD *Energy Prices & Taxes*.

# Changes in Aggregate Energy Intensity

The energy intensities of trucks, ships and rail vary significantly, with trucks being the most intensive (Figure 7.9). On average, trucks use between two and 18 times more energy than rail to move one tonne of goods a distance of one kilometre.

The large range for the energy intensity of truck freight reflects a number of factors. These factors include the type of goods moved, and the size and geography of the country, as well as the split between urban delivery trucks and long-haul trucks, which are much larger and less energy intensive. Given the importance of long-distance haulage in Australia and the United States, it is not surprising that these countries have lower than average energy intensities for freight trucking. Austria, Germany and Sweden also have low truck freight energy intensities. Smaller

7

**Figure 7.9 ▶** *Freight Transport Energy Use per Tonne-kilometre by Mode and Country, 2004*

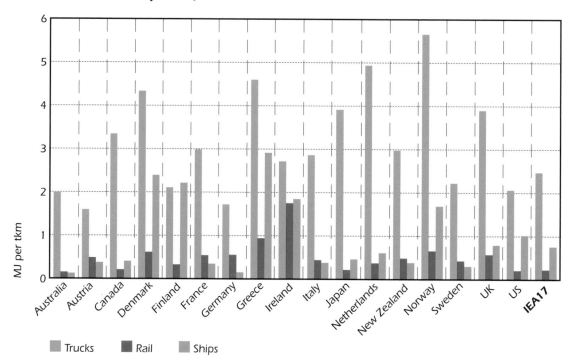

Trucks    Rail    Ships

countries with high population densities, such as Denmark, Japan and the Netherlands, tend to have higher energy intensities for truck freight, reflecting the fact that they have a larger share of short-haul and delivery freight activity and relatively little long-haul trucking. Smaller hilly or mountainous countries, such as Greece, New Zealand and Norway, also tend to have higher truck energy intensities. In some countries, congestion undoubtedly also contributes to higher energy intensities for freight trucks.

Rail energy intensities vary by a factor of six in the IEA17 countries, although this is reduced to a factor of four if Ireland and Greece are excluded. Freight shipping energy intensities vary by a factor of 22, with Denmark, Finland and Greece having particularly high intensities due to a high proportion of short-haul trips. With the exception of Ireland, rail freight has a much lower energy intensity than truck haulage. Shipping also has a lower energy intensity than trucks in all countries except Finland.

The difference in the energy intensity among modes has some important implications for trends in freight energy consumption. First, because of its much higher energy intensity, growth in road freight haulage will have a more significant impact on energy use than growth in freight transport by rail or shipping. Second, intensity reductions in trucking will result in higher energy savings than intensity reductions in rail and shipping – or, indeed, than modal switching between rail and shipping. This effect is amplified by the fact that the share of rail and shipping in freight transport energy use is already relatively small.

These effects help shape the patterns of aggregate energy intensity of freight haulage (as measured by final energy use per tonne-kilometre) across IEA countries. Aggregate energy intensity varies by more than a factor of four; within this range, countries lie within three distinct bands (Figure 7.10). The large differences reflect many factors, but particularly the relative importance of trucking versus rail freight. Countries with the lowest intensities have high shares of rail freight transport (notably Australia, Canada and the United States). Conversely, the highest energy intensities per tonne-kilometre are generally found in smaller countries that have low shares of rail freight (*e.g.* Denmark, Greece and Norway).

**Figure 7.10 ▶** *Freight Transport Energy per Tonne-kilometre Aggregated for All Modes*

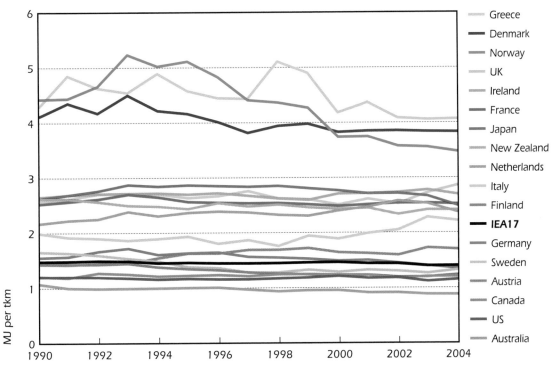

The aggregate freight energy intensity for the IEA17 has been remarkably stable over time, showing only a slight decline (0.4% per year) between 1990 and 2004. Trends in the aggregate energy intensity of freight haulage reflect the balance of two opposing trends. On one hand, there has been a steady decline in the energy intensity of individual modes (trucking, rail and shipping) over time. On the other hand, the growing share of trucks (with their higher energy intensity) in the modal mix has pushed the intensity higher. On balance, energy intensity reductions by individual modes have more than offset the switch to more energy-intensive freight trucking for the IEA17 between 1990 and 2004.

A more detailed analysis of individual countries shows that the trend in total freight energy intensity per tonne-kilometre has varied between an increase of 1.2% per year and a decline of 1.7% per year. Most countries have experienced a decline in the energy intensity per tonne-kilometre between 1990 and 2004, with notable exceptions being Finland, Italy, the Netherlands and the United Kingdom. Finland experienced an increase in the energy intensity of both truck and water-borne freight. In Italy, the Netherlands and the United Kingdom, the increase was due to the increased energy intensity of truck freight haulage.

7

Trends in the energy intensity of trucking are particularly important (Figure 7.11). A steady decline – averaging 0.8% per year – is evident in the average energy intensity of truck freight haulage in the IEA17 between 1990 and 2004. Significant reductions were seen in Australia, Austria, Canada, Germany, Japan, New Zealand, Norway, Sweden and the United States. Conversely, Finland, Italy and the United Kingdom experienced increases in the energy intensity of truck freight haulage during the same period. A number of factors affect the average energy intensity of truck freight haulage: the load factor (average load per vehicle); the share of short-haul freight; vehicle fuel efficiency; driving behaviour; traffic congestion; maximum allowable truck weight; and the availability and quality of the infrastructure for freight transport.

**Figure 7.11** ▶ *Energy Intensity for Trucks*

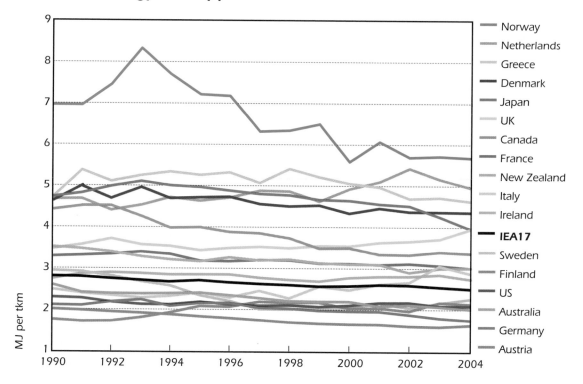

Analysis of the changes in truck load factor for selected countries between 1990 and 2004 (Figure 7.12) can help explain trends in truck energy intensity per tonne-kilometre. Increased weight does contribute to increased fuel consumption, but the marginal increase is not proportionate. Consequently, there is a strong correlation between changes in load factors and changes in the energy intensity of truck freight haulage. Significant improvements in load factors help to explain the reduced energy intensity of truck freight haulage in Australia, Canada, Japan, New Zealand and Norway. In contrast, lower load factors in Italy, the Netherlands and the United Kingdom account for the increase in their energy intensities. Denmark, Sweden and the United States were exceptions: despite lower load factors, they still managed to reduce the energy intensity of truck freight haulage.

**Figure 7.12** ▶ *Truck Average Load per Vehicle*

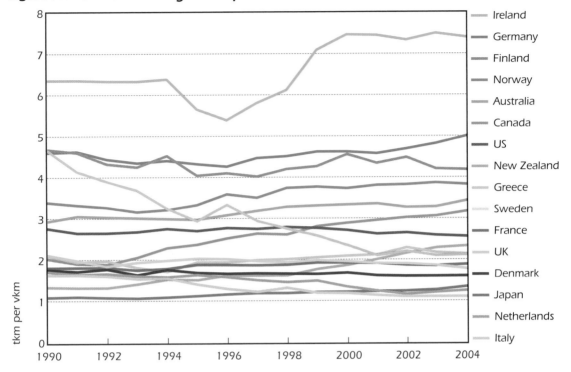

## Impact of Changes in Structure and Intensity

Figure 7.13 shows how changes in total freight energy use are decomposed into changes in total freight haulage (activity), modal structure and energy intensities of the three freight modes included in this analysis (see Annex A). Overall, total energy use in freight transport in the IEA17 increased at an average annual rate of 1.6% between 1990 and 2004. This was a slightly slower growth than the increase in activity, as measured by total tonne-kilometres. Changes in the modal structure – particularly a shift toward more trucking – have tended to increase energy use, but this has been more than offset by reductions in energy intensity, which declined on average by 0.6% per year. However, this rate was slightly lower than in the past; between 1973 and 1990, the rate of energy intensity decline was 0.7% per year (see Annex C).

A slightly different picture emerges when these trends are further separated into specific time periods. Between 1990 and 1995, freight haulage increased rapidly, averaging 2.2% per year. In this period, the reductions in energy intensity were not sufficient to offset the shifts to more intensive modes. Thus, energy use in freight transport increased more rapidly than the level of freight transport activity. Despite rapid economic growth in the second half of the 1990s, increases in freight haulage slowed. The decreases in energy intensities were also smaller, but nonetheless sufficient to outweigh the impact of changes in modal structure. During this period, energy use increased less quickly than activity. In the most recent period, a similar

7

pattern is repeated except that the reductions in intensity are much larger at 1.1% per year, which resulted in a much stronger decoupling of energy use and freight transport activity. Indeed, the intensity reductions from 2000 to 2004 were higher than the average rate over the last 30 years.

**Figure 7.13 ▶** *Factors Affecting Freight Transport Energy Use, IEA17*

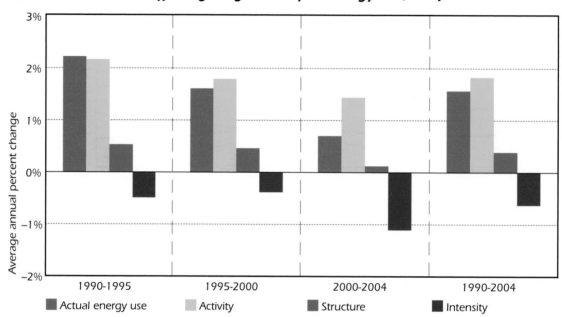

Figure 7.14 presents the decomposition of truck energy intensity, taking account of changes in load factor (the average amount of freight carried by each truck) and in vehicle energy intensity (the average energy use by each truck per vehicle-kilometre in moving freight). For many countries shown, vehicle energy intensity decreased over the period 1990 to 2004, illustrating that, on average, the trucks became individually more efficient. However, the overall energy intensity of trucking was most strongly influenced by the evolution of load factors. For Australia, Canada, France, Germany, Japan, New Zealand and Norway, an increase in load factors (*i.e.* a decrease in vehicle-kilometres per tonne-kilometre) led to a decline in truck energy intensity (measured as truck energy per tonne-kilometre). Conversely, in Finland, Italy, the Netherlands and the United Kingdom, load factors have fallen and the energy intensity of trucking has increased. In Denmark, Sweden and the United States, changes in vehicle energy intensity had a greater impact on trucking energy intensity than did the evolution of load factors.

$CO_2$ emissions have followed a similar evolution to energy use (Figure 7.15) – *i.e.* changes in the fuel mix and the carbon intensity of the fuel mix have had very little impact. This reflects the overwhelming dominance of oil for both trucking and shipping, as compared to the modest use of alternative fuels (*e.g.* biofuels) in these modes and to the relatively minor role of electricity in rail freight. Overall, between 1990 and 2004, the IEA17 countries saw $CO_2$ emissions from freight transport increase at 1.5% per year, very slightly less than the rate of increase for energy.

Separating the period into three sub-periods reveals an increased reduction, albeit small, in the carbon intensity of fuels in the freight transport sector during the period 2000 – 04. This is due, in part, to the growth in biofuels and other oil substitutes. Current biofuels targets in Europe and North America could reduce carbon intensity even further in the future.

**Figure 7.14** ▶ *Decomposition of Changes in Trucks Energy Intensity, 1990 – 2004*

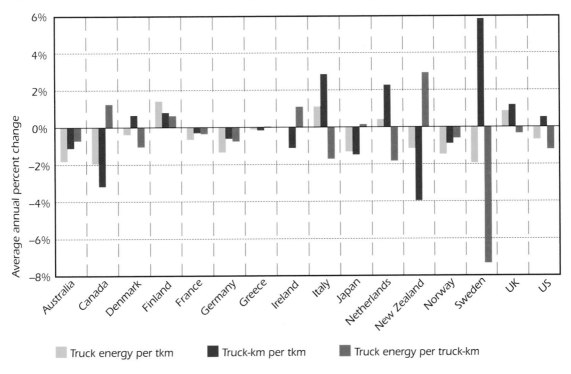

**Figure 7.15** ▶ *Decomposition of Changes in Freight CO$_2$ Emissions, IEA17*

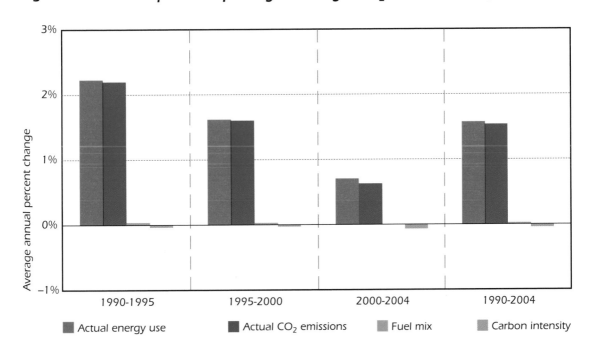

# Energy and CO₂ Savings

The decline in energy intensity of freight transport between 1990 and 2004 has led to savings in both energy and $CO_2$ emissions (Figure 7.16). Annual energy savings were 1.1 EJ in 2004, equivalent to 9% of freight transport energy use in that year. The reduction in $CO_2$ emissions in 2004 was 80 Mt.

**Figure 7.16** ▶ *Freight Transport Energy and CO₂ Emissions Savings from Reductions in Energy Intensity, IEA17*

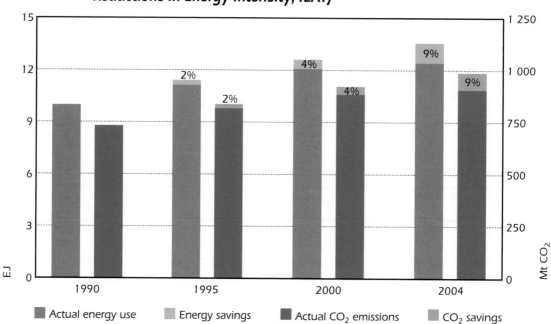

■ Actual energy use     ■ Energy savings     ■ Actual CO₂ emissions     ■ CO₂ savings

# Summary

Examination of domestic freight transport (excluding air freight) in 17 IEA countries over the period 1990 to 2004 reveals the following key findings:

▷ Final energy consumption in freight transport increased by 25%; $CO_2$ emissions followed a similar pattern with a 24% rise. These trends were driven by rising freight haulage (measured in tonne-kilometres), which increased by 32%, and by the increased share of trucking.

▷ Trucks are responsible for most of the energy use in freight transport, with a share of 82% in 2004. There are significant differences in the modal split from country to country. Rail and shipping both have large shares of total freight haulage in Australia, Canada and the United States. Rail is also important in Austria and Sweden. Shipping has a significant share in Greece, Japan, the Netherlands and Norway.

▷ Total freight haulage has increased on a per capita basis in many IEA countries. However, the overall trend per unit of GDP has been slightly downward. Australia, Canada and the United States are the most freight-intensive countries per unit of GDP, reflecting their large size and the importance of raw materials production and transport. However, these countries also have the lowest freight energy use per tonne-kilometre, due to the significant share of freight haulage by rail.

▷ Trucking energy intensity has declined significantly, at 0.8% per year. The most notable reductions were achieved in Australia, Austria, Canada, Germany, Japan, New Zealand, Norway and Sweden. Several factors contributed to these reductions including higher load factors coupled with improvements in vehicle efficiencies. In some cases, reductions can be attributed to changes in driver behaviour and the increased share of large, long-haul freight trucking that carry heavy loads very efficiently.

▷ The reduced energy intensity of trucking was the main driver behind the decline in overall energy intensity of freight transport, which fell by an average rate of 0.6% per year. However, this rate of reduction was slightly lower than in previous decades; a fall of 0.7% per year was seen between 1973 and 1990.

▷ With these reductions in energy intensity, energy use in 2004 was 1.1 EJ or 9% lower than it would have been otherwise; $CO_2$ emissions were 80 Mt lower.

7

# CONCLUSIONS AND POLICY IMPLICATIONS

## Trends in Energy Use and $CO_2$ Emissions

Since 1990, improvements in energy efficiency have continued to play a key role in shaping energy use and $CO_2$ emissions patterns in IEA countries. By 2004, these improvements had led to an annual energy saving of 16 EJ, which is equivalent to 1.2 Gt of avoided $CO_2$ emissions and an estimated USD 170 billion of energy cost savings. However, it is clear that more could be done; gains in energy efficiency have been less than 1% per year since 1990; significantly lower than in previous decades.

The results of this study confirm the conclusion of the IEA's previous indicator publication (IEA, 2004) – *i.e.* that the changes caused by the oil price shocks in the 1970s and the resulting energy policies did considerably more to control growth in energy demand and reduce $CO_2$ emissions than the energy-efficiency and climate policies implemented since the 1990s.

The current low rate of energy-efficiency improvement poses a substantial challenge to IEA countries in their efforts to ensure reliable, affordable and clean energy for their citizens. Two trends are a particular cause for concern. The first is the rapid increase in electricity consumption, driven largely by higher demand from the household and service sectors. In many IEA countries, electricity production is primarily based on fossil fuels; thus, increased electricity consumption is a major contributor to rising $CO_2$ emissions. The second concern stems from the continuing growth of passenger and freight transport activity and from the low rate of overall improvement in energy efficiency, despite better vehicle and engine technologies. Together, passenger and freight transport now account for over 75% of oil use by final consumers. Despite recent efforts in IEA countries to increase the share of alternatives to gasoline and diesel, the transport sector is likely to remain heavily reliant on oil products for many years to come.

## Implications for Energy-Efficiency Policy

Projections published by the IEA in *Energy Technology Perspectives 2006* clearly demonstrate the need to double the recent rate of improvement in energy efficiency to have a realistic chance of a more sustainable energy future. The good news is that this is indeed possible; there is still significant scope in IEA countries for energy-efficiency improvements in buildings, industry and transport. However, a key lesson of the past is that the widespread future deployment of more efficient technologies and practices will require very strong government action. The challenge is to find the right mix of market- and regulation-based policies to achieve cost-effective decisions regarding energy efficiency in all sectors. This should be complemented by efforts to drive down the $CO_2$ intensity of electricity production by moving towards a cleaner technology mix.

Based on this analysis and other work on best practices in policies and measures, the IEA has identified the following key areas in which governments need to take urgent action.

▷ Appliances and equipment, particularly those using electricity, represent one of the fastest growing energy demands in most IEA countries. Governments should take steps to introduce mandatory energy performance standards and, where appropriate, energy labelling across the full range of mass-produced equipment. Action is also needed to make it a standard feature for individual and networked devices to switch automatically to low-power modes when not in use.

▷ To combat the effects of fast growing energy use in transport and the corresponding $CO_2$ emissions, countries should implement mandatory fuel efficiency standards for cars and small trucks. They should also adopt international test procedures for measuring tyre-rolling resistance.

▷ Buildings account for almost 40% of energy used in IEA countries. To save a significant portion of this energy, governments should strengthen the energy-efficiency requirements of building codes and promote low-energy houses. Systematic effort is also needed to monitor energy-efficiency improvements in existing buildings.

▷ Saving energy by adopting efficient lighting technology is particularly cost-effective. Governments should take steps to phase out the most inefficient incandescent bulbs as soon as commercially and economically viable.

To make these suggested policies a reality, governments will need to provide adequate resources (both financial and human) to the policy agencies responsible for energy efficiency. This should include funding and managing the implementation and monitoring of action plans for energy efficiency. In turn, this will require the establishment and ongoing maintenance of high quality energy indicators and other data.

# The Role of Indicators in Policy Development and Evaluation

Collecting data carries a cost for companies, organisations and countries. However, lack of data – or not having the right data – may lead to misinformed policy decisions and sub-optimal choices that prove to be far more costly in the long term. Thus, it is important to ask what data are needed to understand patterns of energy use and $CO_2$ emissions, and to track progress regarding improvements in energy efficiency.

At a minimum, all countries should collect annually the set of energy statistics needed to build an energy balance. These statistics can then be coupled with basic economic statistics (such as GDP and sectoral value-added) to calculate aggregate energy intensity indicators (such as the ratio of total final energy consumption to GDP). Still, such indicators are not sufficient for understanding the driving forces behind the changing patterns of energy consumption or for examining the role of energy efficiency in the economy. Additional energy and socio-economic data are needed to focus on a particular sector and to isolate the underlying influences and trends.

IEA countries have already recognised the importance of moving beyond energy balances and have set up frameworks to collect detailed statistics in various sectors. In Europe, the ODYSSEE project, which is funded by the European Commission and country governments, is a notable example of such a mechanism. Countries in other IEA regions have also established data collection processes for specific sectors.

The availability of detailed end-use data has increased in the past few years. This publication has benefited from the efforts made by IEA countries to collect and release more and better statistics. Consequently, for each of 20 IEA countries it has been possible to analyse data for at least two sectors, compared to the 14 countries covered in the indicator analysis published in 2004. This illustrates that countries recognise the growing importance of indicators work to support the development and evaluation of energy-efficiency policies.

Despite this progress, IEA countries still need to address a number of areas relating to completeness, timeliness and quality of the information:

▷   Completeness: Approximately half of IEA countries are not yet in a position to provide some or all of the data needed for the indicators presented in this analysis.

▷   Timeliness: The main concerns in this area are delays in making data available and time lags between consecutive surveys needed to support the development of such data.

▷   Data quality: This broad term encompasses issues of comparability and reliability. In terms of comparability, discrepancies can be observed between the disaggregated sectoral data (collected by the IEA to support the indicators work) and the aggregated data (provided by countries for the annual IEA statistics questionnaires and used to prepare national energy balances). This reflects the fact that, in some cases, sources of detailed data for indicators are different from official sources. Regarding reliability, it is possible to observe anomalies in totals, breaks in time series, missing data, etc. These inconsistencies reduce the reliability and usefulness of the data.

This last point raises the question of using official government data versus data collected from a variety of other sources. This issue is not specific to indicators but applies to any kind of statistics. However, as indicators can be used to make comparisons across countries, it is particularly important that this is based on similar data definitions and methodologies, which have been approved by the countries. The ODYSSEE indicators programme establishes a framework, guidelines and methodologies to encourage such harmonisation among European countries. The IEA extends the concept of harmonisation to other parts of the world. For example, in preparing this publication, the IEA developed a set of templates to facilitate data compilation according to a well-structured design and definitions. However, the process is still some way from providing an ideal mechanism for the timely submission of comprehensive and complete data in a common format.

Several non-IEA countries have embarked upon programmes to develop indicators that reflect their own situations. This is extremely encouraging, especially at a time when policy makers in many parts of the world are making energy efficiency a priority. The IEA is already working with a number of non-member countries and international institutions on energy statistics and indicators activities, and is keen to share its knowledge and experience.

This publication demonstrates the ways in which high quality, comprehensive and timely energy data and indicators can act as key tools for analysing energy use and efficiency developments. It also seeks to help governments recognise the critical importance of statistical and indicator activities. In this regard, the IEA wishes to emphasise the need for governments to realistically assess the scope of the work involved, and to allocate sufficient resources. Failure to do so will diminish the usefulness of the information collected – and will ultimately undermine the ability of analysts and policy makers to develop, implement and monitor successful energy-efficiency policies.

# ANNEX A: METHODOLOGY AND DATA SOURCES

## The IEA Methodology for Analysing Energy Use

The methodology used in this study builds on the analytical framework presented in *Oil Crises and Climate Challenges: 30 Years of Energy Use in IEA Countries* (IEA, 2004).

The analysis of energy end-use trends distinguishes between three main components affecting energy consumption: aggregate activity, sectoral structure and energy intensities (see Table A.1 for more details).

▷ **Aggregate activity (A)** is measured in one of the following ways, depending on the sector: as value-added for manufacturing industry and services; as population in the household sector; or as passenger-kilometres and tonne-kilometres, respectively, for the passenger and freight transport sectors.

▷ **Sectoral structure (S)** represents the mix of activities within a sector and further divides activity into industry sub-sectors, measures of residential end-use activity or transportation modes.

▷ **Energy intensity (I)** refers to energy use per unit of activity.

To separate the effect of various components over time, the IEA uses a factoral decomposition approach that analyses changes in energy use within a sector, using the following equation:

$$E = A \cdot \sum_{r}(S^r \cdot I^r)$$

In this decomposition, the symbols represent the following parameters:

E    Total energy use in a sector.

A    Overall sectoral activity.

r    Sub-sectors or end-uses within a given sector.

$S^r$    Share of sub-sector or end-use "r" in a sector.

$I^r$    Energy intensity of each sub-sector or end-use "r".

The **activity effect** can be calculated as the relative impact on energy use that would have occurred in year $t$ if the structure and energy intensities for a sector had remained fixed at their base year values (t=0) while aggregate activity had followed its actual development.

$$E_t^A = \frac{A_t \cdot \sum_{r}(S_0^r \cdot I_0^r)}{E_0}$$

**Table A.1 ▶ *Summary of Variables Used***

| Sector | Sub-sector | Activity (A) | Structure (S) | Intensity (I) |
|---|---|---|---|---|
| **Households** | | | | |
| | Space Heating | Population | Floor Area/Population | Space Heating Energy[1] /Floor Area |
| | Water Heating | " | Population/ Occupied Dwellings | Water Heating Energy[2] /Occupied Dwellings |
| | Cooking | " | Population/ Occupied Dwellings | Cooking Energy[2] / Occupied Dwellings |
| | Lighting | " | Floor Area/Population | Lighting Energy /Floor Area |
| | Appliances | " | Appliances Ownership / Population | Appliances Energy / Appliances Ownership |
| **Passenger Transport** | | | | |
| | Car | Passenger-kilometre | Share of Pass-kilometre | Energy/Pass-kilometre |
| | Bus | " | " | " |
| | Rail | " | " | " |
| | Domestic Air | " | " | " |
| **Freight Transport** | | | | |
| | Truck | Tonne-kilometre | Share of Tonne-kilometre | Energy/Tonne-kilometre |
| | Rail | " | " | " |
| | Domestic Shipping | " | " | " |
| **Manufacturing** | | | | |
| ISIC 15 – 16 | Food, Beverages & Tobacco | Value-added | Share of Value-added | Energy/Value-added |
| ISIC 21 – 22 | Paper, Pulp & Printing | " | " | " |
| ISIC 24 | Chemicals | " | " | " |
| ISIC 26 | Non-metallic Minerals | " | " | " |
| ISIC 27 | Primary Metals | " | " | " |
| ISIC 28 – 32 | Metal Products & Equipment | " | " | " |
| ISIC 17 – 20, 25, 33 – 37 | Other Manufacturing | " | " | " |
| **Services** | | | | |
| ISIC 50 – 99 | Services | Value-added | Share of Value-added | Energy/Value-added |
| **Other Industries[3]** | | | | |
| ISIC 1 – 5 | Agriculture & Fishing | Value-added | Share of Value-added | Energy/Value-added |
| ISIC 45 | Construction | " | " | " |

[1] Adjusted for climate variations using heating degree-days.
[2] Adjusted for household occupancy.
[3] The following ISIC groups are not included in the analysis: 10 – 14 Mining & Quarrying; 23 Fuel Processing; and 40 – 41 Electricity, Gas & Water Supply. Industries in category "Other industries" are analysed only to a very limited extent in this study.

Similarly, the **structure effect** is determined by making the calculation using constant aggregate activity and energy intensities but varying the sectoral structure.

$$E_t^s = \frac{A_0 \cdot \sum_r (S_t^r \cdot I_0^r)}{E_0}$$

The **intensity effect** is calculated by assuming that the sectoral structure and aggregate activity for a sector had remained fixed at the base year values while energy intensities had followed their actual development.

$$E_t^I = \frac{A_0 \cdot \sum_r (S_0^r \cdot I_t^r)}{E_0}$$

Thus, by calculating the relative impact on energy use from changes in each of these components, it is possible to isolate the impacts on energy use related to improved end-use energy efficiency (reductions in energy intensities) – *i.e.* to separate them from changes deriving from shifts in the activity and structure components.

In this analysis, the **hypothetical energy use** (HEU$^I$) is defined as the energy use that would have occurred in year t if energy intensities in each sector remained constant at their base year values. It is calculated by dividing actual energy use in year t by the intensity effect in that year.

$$HEU_t^I = \frac{E_t}{E_t^I}$$

**Energy savings** from reduced energy intensities can be defined as the difference between the hypothetical energy use and actual energy use.

$$SAVINGS_t^I = HEU_t^I - E_t$$

By introducing the dimension of fuel mix and carbon intensity (or $CO_2$ intensity), the decomposition of energy use can be extended to address changes in $CO_2$ emissions (G). In this case, fuel mix (F) represents changes in fuel shares (including electricity) among end-uses. Carbon intensity (C) refers to the $CO_2$ emissions per unit of energy used.

$$F_t^{r,f} = \frac{E_t^{r,f}}{E_t^r} \qquad\qquad C_t^{r,f} = \frac{G_t^{r,f}}{E_t^{r,f}}$$

The **$CO_2$ emissions** (G) in a sector can then be decomposed into the activity, structure, energy intensity, fuel mix and carbon intensity effects according to the following formula:

$$G_t = A_t \cdot \sum_r \left[ S_t^r \cdot I_t^r \cdot \sum_{f \in fuel} \left( F_t^{r,f} \cdot C_t^{r,f} \right) \right]$$

This makes it possible to calculate the hypothetical $CO_2$ emissions as well as $CO_2$ savings. For example, the following two formulas present the carbon intensity effect and corresponding savings.

$$G_t^C = \frac{A_0 \cdot \sum_r \left[ S_0^r \cdot I_0^r \cdot \sum_f \left( F_0^{r,f} \cdot C_t^{r,f} \right) \right]}{G_0} \qquad\qquad CO_2\ SAVINGS_t^C = \frac{G_t}{G_t^C} - G_t$$

**Figure A.1** ▶ *Basic Overview of Factors in CO$_2$ Decomposition*

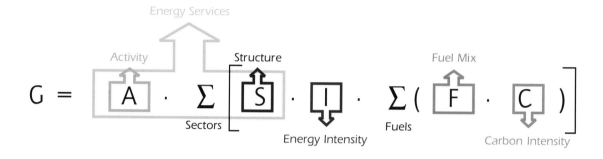

Regional aggregates for hypothetical energy use are calculated as the sum of hypothetical energy uses across all countries in a particular region. Energy savings for a region are then calculated as a difference between the hypothetical energy use and the actual energy use. The same approach is used for CO$_2$ emissions.

A number of different index-number techniques can be used to analyse factors affecting energy use. This book uses the Laspeyres indices approach – one of the most popular techniques, due to the fact that it is relatively simple to interpret. The Laspeyres indices have a second advantage for this analysis in that this method maintains broad consistency with the approach used in previous IEA indicator publications. However, the indicator results are affected by this choice of approach, and by the selection and definition of the activity, structure and intensity variables. In addition, it is important to keep in mind that individual countries that have created their own indicators (such as described in Annex D) may have different results. Some of these countries have used a different decomposition method and chosen different variables.

## Long-term Trends

The calculation of long-term trends (shown in Annex C) was undertaken by combining two analyses for 11 IEA countries:[4]

▷   Data and results published in *Oil Crises and Climate Challenges: 30 years of Energy Use in IEA Countries* (IEA, 2004) were used for the period 1973 to 1989.

▷   Data and results from the current publication were used for the period 1990 to 2004.

Some of the data for 1990 to 1998 may differ from that previously published by the IEA (IEA, 2004) due to revisions or to variations in base year numbers and in deflators for key activity or climate-correction factors. Therefore, the results for the 1973 – 98 period (which are not explicitly presented in this current publication) may differ from those previously published. In addition, there are slight differences in approach and data between the current methodology and that used previously. Thus,

---

4. The IEA11 includes Australia, Denmark, Finland, France, Germany, Italy, Japan, Norway, Sweden, the United Kingdom and the United States, i.e. those countries for which data are available for the whole time period 1973-2004.

the combined decomposition analysis presented in Annex C should be considered indicative of the results that would be obtained if it were possible to conduct a single, complete decomposition for the period 1973 to 2004.

# Sectors and Data Analysed

## *Sectoral Coverage*

This analysis considers energy use in the manufacturing, household, service, passenger and freight transport sectors in the categories shown in Figure A.2. It does not consider "other industries" in detail, as data for these activities are scarce.

All energy data in this publication are expressed on a net calorific value basis (using lower heating values). Data definitions are based on the methodology used in the IEA energy statistics and balances, although there are some important differences. In the IEA energy balances, coal transformation losses are included as energy transformation. The IEA indicator approach allocates these losses to the primary metals sector (ISIC 27) in which the secondary coal products are consumed. With the energy balances method, petroleum products used as feedstocks for industrial chemicals are included as non-energy use in the total final consumption. These products are not included at all in the indicator approach. Similarly, the energy balances approach includes energy use for refining in the transformation sector whereas the indicator approach considers refining as part of ISIC 23 (manufacture of coke, refined petroleum products and nuclear fuels), which is excluded in this study.

This study also excludes some aspects of transportation such as natural gas pipelines, and fuel use for private boats and military vehicles. Both approaches exclude international marine bunkers from total final energy consumption (TFC). International air traffic is included in the IEA statistics but not in the indicator approach. As a result of these differences, TFC for IEA countries in 2004 is about 10% higher using the IEA statistics approach than with the indicator methodology.

Further information on the scope of individual sectors is provided below.

The **manufacturing sector** of industry produces finished goods or products for use by other businesses, for sale to domestic consumers or for export. This publication disaggregates total manufacturing into several key industries.

The **household sector** includes those activities related to private dwellings. It covers all energy-using activities in apartments and houses, including space and water heating, cooking, lighting and the use of appliances.

The **service sector** includes activities related to trade, finance, real estate, public administration, health, education and commercial services.

**Passenger transport** includes the movement of people by road, rail, sea and air. Passenger road transport is further subdivided into cars and buses. International air travel is excluded due to a lack of consistent and comparable data for IEA countries.

**Freight transport** includes the movement of goods by road, rail and sea. It excludes air freight transport because of a lack of comprehensive and consistent data for this mode.

**Figure A.2** ▶ *Disaggregation of Sectors, Sub-sectors, and End-uses in IEA Energy Indicators Approach*

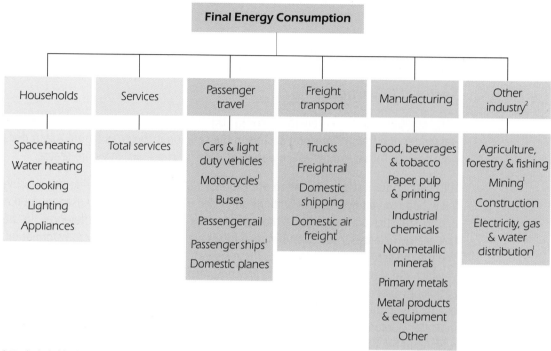

[1] Not included in this study due to lack of consistent and reliable data series.

[2] Other Industry is included in the analysis only in the chapter Overall Trends and is not analysed separately.

### Country Coverage

The types of detailed data that are required for time-series analysis exist in many IEA countries, but not yet in all of them. This study considers energy use in those IEA countries for which consistent, detailed, long-term time series are available for a particular sector. Table A.2 (below) summarises the country coverage in each sector.

## Data Sources

Due to the diverse nature of the data needed for the disaggregated indicator analysis, this study draws from a mix of national and international sources. To maintain consistency across countries, if possible, the data were taken from the following OECD or IEA statistics.

▷ Energy Balances of OECD Countries, 2006, IEA

▷ Energy Prices & Taxes, 4th quarter 2006, IEA

▷ Mobility Modelling Database, IEA

▷ National Accounts of OECD Countries, 2007 Volume 1, OECD

▷ The OECD STAN Database for Industrial Analysis, November 2005, OECD

▷ OECD Main Economic Indicators, January 2007, OECD

▷ OECD Economic Outlook, no. 80, OECD

The IEA also worked closely with the following national and international organisations and groups to obtain detailed energy and activity data for manufacturing, households, passenger travel and freight transport.

▷ ODYSSEE project "Energy Efficiency Indicators", led by ADEME and supported by the EIE programme of the European Commission/DGTREN, or from national teams within this project.[5]

▷ Eurostat Unit G4 "Energy Statistics" and JRC IPSC/Agrifish Unit/MARS-STAT Action.

▷ Office of Energy Efficiency, Natural Resources Canada

▷ Energy Efficiency and Conservation Authority and the Ministry of Economic Development, New Zealand.

▷ Department of Energy (Energy Information Administration and Office of Energy Efficiency and Renewable Energy) and the Department of Transportation (Bureau of Transportation Statistics), United States.

▷ Australian Bureau of Agricultural and Resource Economics, Australian Bureau of Statistics, Australian Greenhouse Office and Bureau of Transportation and Regional Economics.

▷ Ministry of Economy, Trade and Industry and the Energy Data and Modelling Center, Japan.

▷ Department for Business, Enterprise and Regulatory Reform and the Deparment for Environment, Food and Rural Affairs, United Kingdom.

## Table A.2  ▶ *Country Coverage by Sector*

| | Manufacturing & Services | Transport | Households | Economy-wide Trends |
|---|---|---|---|---|
| Australia | ✔ | ✔ | ✗ | ✗ |
| Austria | ✔ | ✔ | ✔ | ✔ |
| Belgium | ✔ | ✗ | ✗ | ✗ |
| Canada | ✔ | ✔ | ✔ | ✔ |
| Denmark | ✔ | ✔ | ✔ | ✔ |
| Finland | ✔ | ✔ | ✔ | ✔ |
| France | ✔ | ✔ | ✔ | ✔ |
| Germany | ✔ | ✔ | ✔ | ✔ |
| Greece | ✔ | ✔ | ✗ | ✗ |
| Ireland | ✗ | ✔ | ✗ | ✗ |
| Italy | ✔ | ✔ | ✔ | ✔ |
| Japan | ✔ | ✔ | ✔ | ✔ |
| Netherlands | ✔ | ✔ | ✔ | ✔ |
| New Zealand | ✔ | ✔ | ✔ | ✔ |
| Norway | ✔ | ✔ | ✔ | ✔ |
| Portugal | ✔ | ✗ | ✗ | ✗ |
| Spain | ✔ | ✗ | ✔ | ✗ |
| Sweden | ✔ | ✔ | ✔ | ✔ |
| United Kingdom | ✔ | ✔ | ✔ | ✔ |
| United States | ✔ | ✔ | ✔ | ✔ |
| | IEA19 | IEA17 | IEA15 | IEA14 |

5. The ODYSSEE national teams in the European countries covered in this study include; ADEME (France), ADENE (Portugal), AEA (Austria), AEA Technology (United Kingdom), DEA (Denmark), ECN (Netherlands), Econotec (Belgium), ENEA (Italy), FhG-ISI (Germany), IDAE (Spain), IFE (Norway), MOTIVA (Finland), SEI (Ireland) and STEM and Statistics Sweden (Sweden).

# ANNEX B: SELECTED INDICATORS USING GDP AND VALUE-ADDED AT MARKET EXCHANGES RATES

This Annex shows some of the key energy and $CO_2$ indicators from the main chapters, re-calculated using GDP or value-added at market exchange rates (MER), rather than at purchasing power parities (PPP). Box 2.2 in Chapter 2 explains more about these two different approaches.

**Figure B.1** ▶ *Energy Use per Manufacturing Value-added*

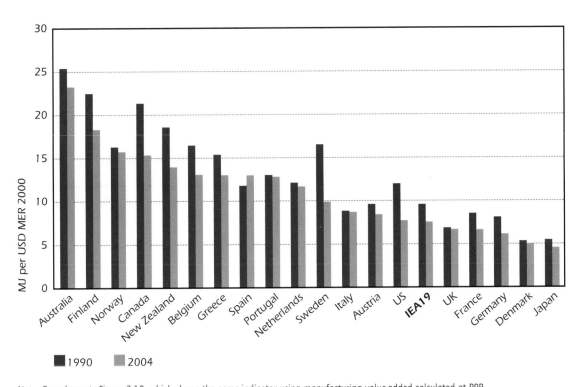

Note: Complements Figure 3.10, which shows the same indicator using manufacturing value-added calculated at PPP.

**Figure B.2** ▶ *Manufacturing Energy Intensity at a Common IEA19 Structure*

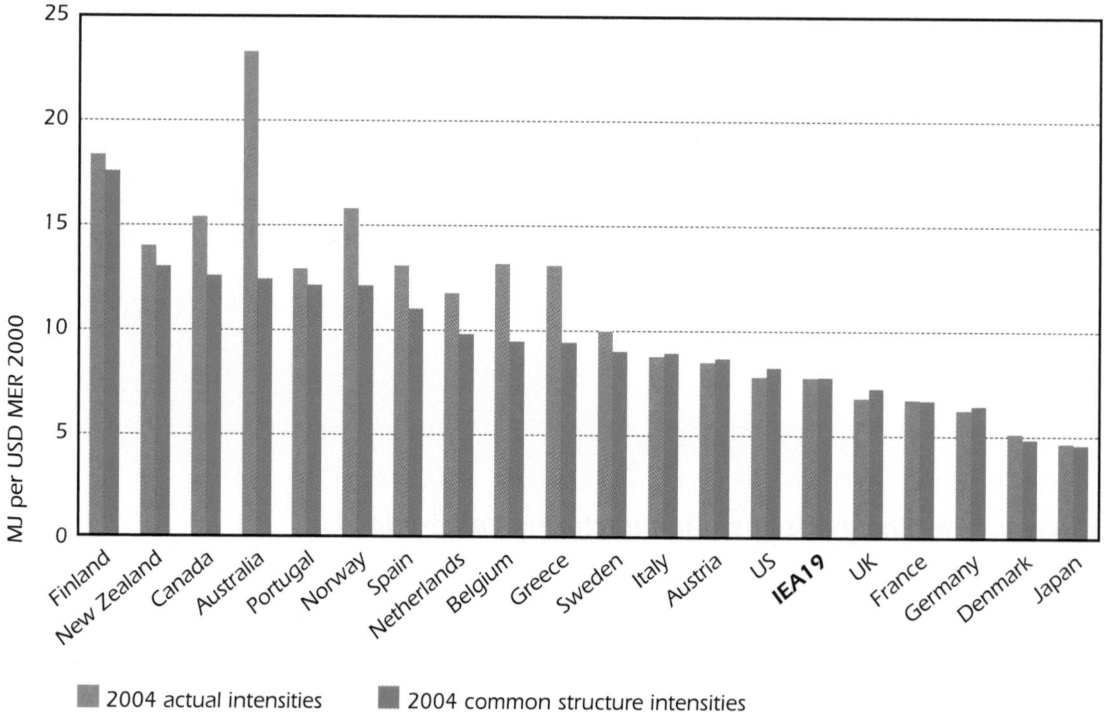

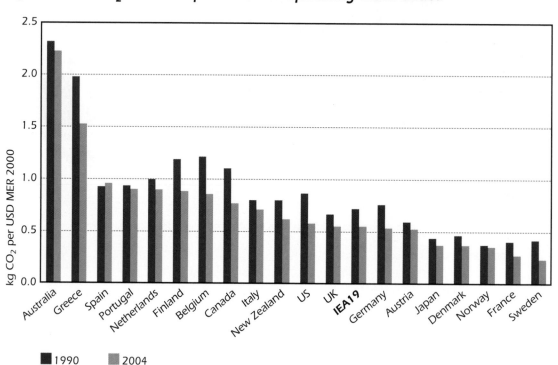

■ 2004 actual intensities    ■ 2004 common structure intensities

Note: Complements Figure 3.12, which shows the same indicator using manufacturing value-added calculated at PPP.

**Figure B.3** ▶ *CO$_2$ Emissions per Total Manufacturing Value-added*

■ 1990    ■ 2004

Note: Complements Figure 3.13, which shows the same indicator using manufacturing value-added calculated at PPP.

**Figure B.4** ▶ *Service Sector Electricity Use per Capita and Value-added per Capita*

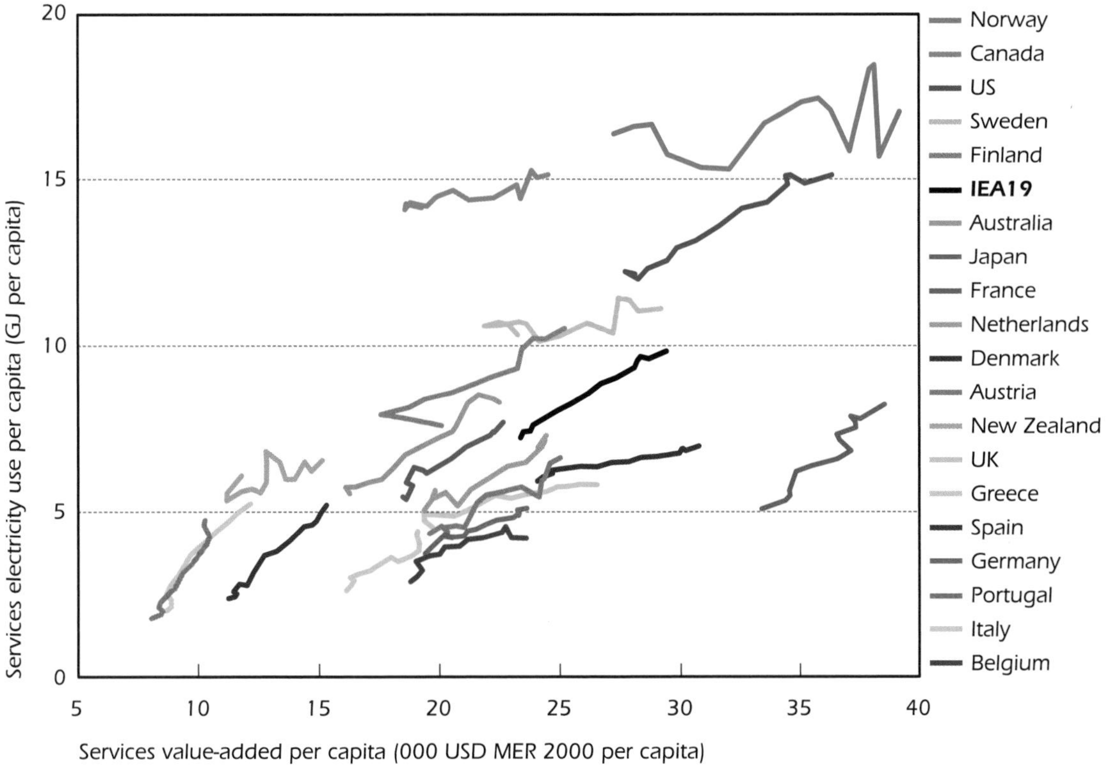

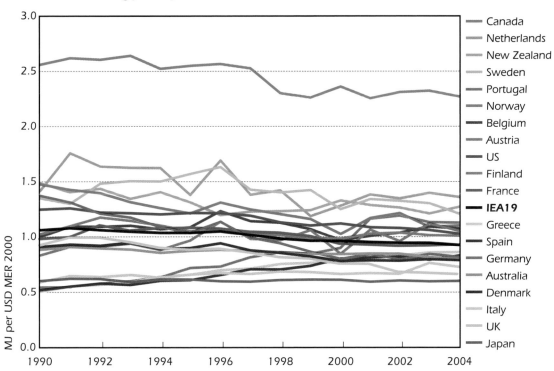

Note: Complements Figure 5.6, which shows the same indicator using service value-added calculated at PPP.

**Figure B.5** ▶ *Energy Use per Service Sector Value-added*

Note: Complements Figure 5.7, which shows the same indicator using service value-added calculated at PPP.

**Figure B.6** ▶ *Freight CO₂ Emissions per GDP*

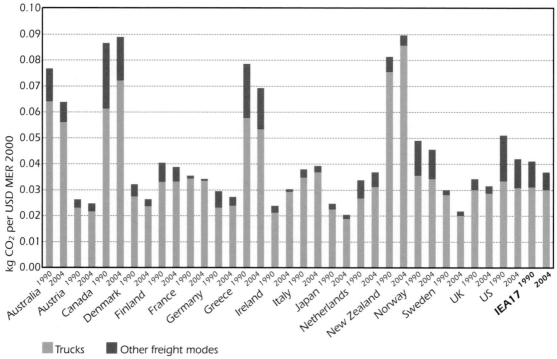

■ Trucks   ■ Other freight modes

Note: Complements Figure 7.4, which shows the same indicator using GDP calculated at PPP.

**Figure B.7** ▶ *Total Freight Tonne-Kilometres per GDP, All Modes*

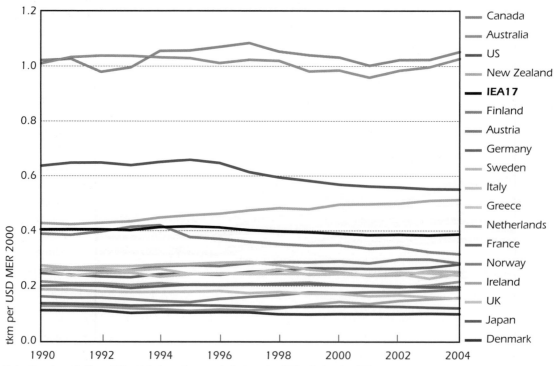

Note: Complements Figure 7.5, which shows the same indicator using GDP calculated at PPP.

# ANNEX C: LONG-TERM TRENDS

This publication focuses on patterns of energy use and $CO_2$ emissions during the period from 1990 to 2004. However, it is interesting to put these in the context of longer term trends by revisiting the results of a previous IEA indicator publication (IEA, 2004) that examined the period from 1973 to 1998 for a group of 11 countries.[1] Annex A explains how the long-term trends have been constructed and describes some of the problems in achieving a completely consistent dataset over such a long period of time. Consequently, the results presented in this Annex should be considered only as indicative of the longer term impacts of energy efficiency.

Without the energy efficiency improvements achieved since 1973, energy use for the 11 IEA countries would have been 56% higher in 2004 than it actually was, the equivalent of 58 EJ of energy not consumed (Figure C.1). This makes energy savings the most important "fuel" in the IEA11 for this time period – i.e. the amount of energy saved in 2004 was slightly higher than the actual consumption of oil, or of electricity and gas combined. By comparison, the IEA11 achieved energy savings of 38% in 1990 and 49% in 1998, due to improvements in energy efficiency since 1973. Despite impressive contributions from energy efficiency, the rate of energy savings has slowed considerably since 1990. Between 1973 and 1990, the average rate of improvement in energy efficiency was close to 2%; after 1990, the rate dropped by more than one-half to 0.9% per year.

Energy consumption trends over the same time period show a stark contrast: from 1990 to 2004, energy use grew at twice the rate recorded between 1973 and 1990. Such a trend is a serious challenge in terms of environmental protection and energy security. These results reaffirm a key conclusion of the earlier indicators publication – namely, that the oil price shocks of the 1970s and the resulting energy policies did considerably more to control growth in energy demand and $CO_2$ emissions than the energy-efficiency and climate policies implemented in the 1990s.

A similar pattern emerges from analysis of the individual end-use sectors, as shown in Figures C.2-C.6. Energy efficiency improvements have led to significant savings in all sectors. However, without exception, the rate of improvement since 1990 has been lower than in the period between 1973 and 1990. In manufacturing, the strong energy-efficiency improvements between 1973 and 1990 led to a decrease in energy use, but this reduction did not continue into the 1990s. For households and services, lower energy-efficiency improvements since 1990 have led to significantly higher increases in energy use between 1990 and 2004 than in earlier decades. For both passenger and freight, the reduction in the rate of improvements in energy efficiency since 1990 has been partially compensated by lower increases in energy service demand. For passenger transport, the rate of increase in energy use changed little between the two periods. The rate of increase in energy use for freight transport actually fell after 1990.

---

1. The 11 IEA countries included in the analysis of the long-term trends are Australia, Denmark, Finland, France, Germany, Italy, Japan, Norway, Sweden, the United Kingdom and the United States.

**Figure C.1** ▶ *Long-term Energy Savings from Improvements in Energy Efficiency, All Sectors, IEA11*

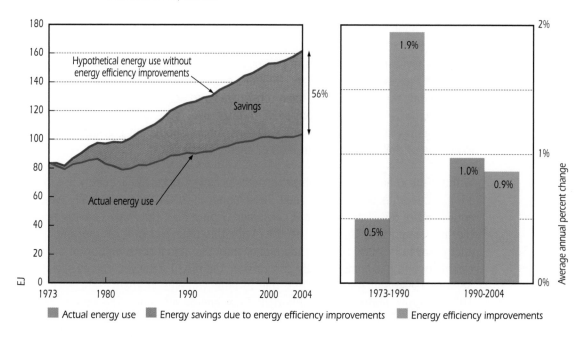

**Figure C.2** ▶ *Long-term Energy Savings from Improvements in Energy Efficiency, Manufacturing, IEA11*

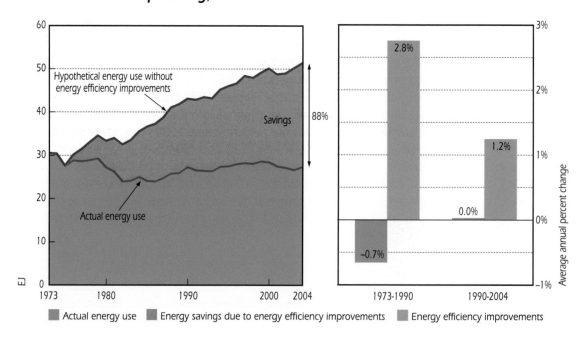

**Figure C.3** ▶ *Long-term Energy Savings from Improvements in Energy Efficiency,*
*Households, IEA11*

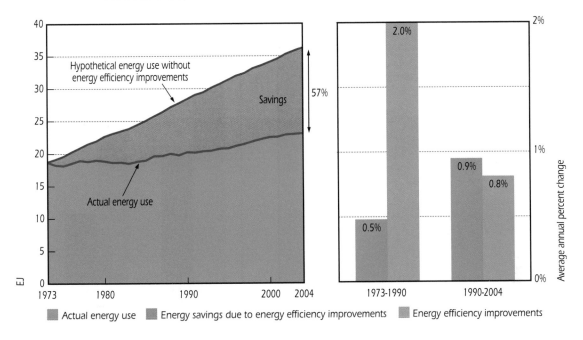

**Figure C.4** ▶ *Long-term Energy Savings from Improvements in Energy Efficiency,*
*Services, IEA11*

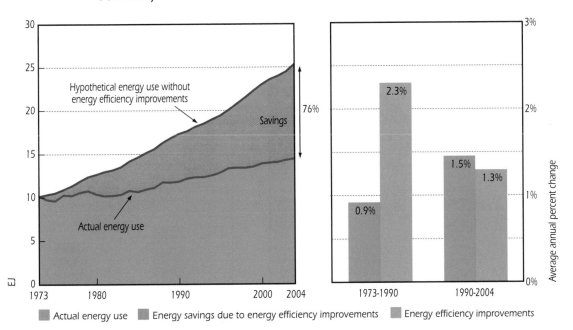

**Figure C.5** ▶ *Long-term Energy Savings from Improvements in Energy Efficiency, Passenger Transport, IEA11*

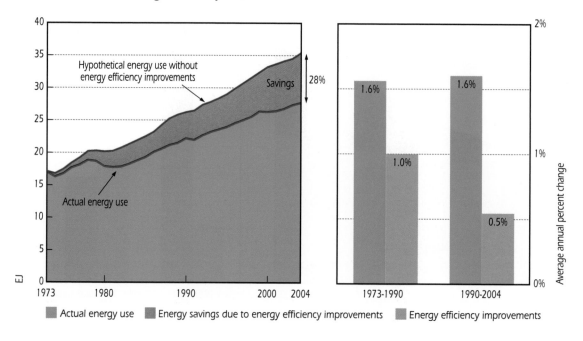

**Figure C.6** ▶ *Long-term Energy Savings from Improvements in Energy Efficiency, Freight Transport, IEA11*

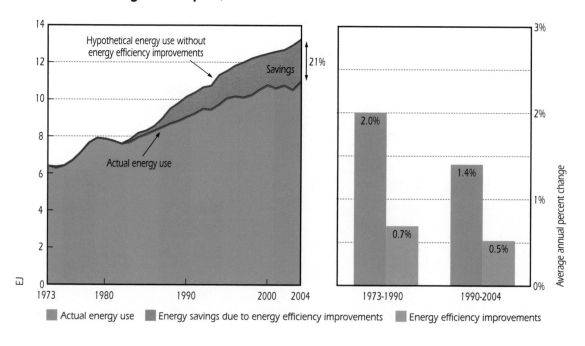

# ANNEX D: APPLICATIONS OF ENERGY INDICATORS IN IEA COUNTRIES

Many IEA countries already use energy indicators as a tool for policy making; several of these countries publish energy indicator studies or books. The following paragraphs summarise the current status of energy indicators work in each country.

## Australia

Australia is developing a framework for energy-efficiency data to facilitate better understanding of energy-efficiency initiatives across the economy. One of the key challenges is how to improve the availability of data on Australia's building stocks and their energy performance, especially in relation to water heating and insulation.

Australia's Department of Industry, Tourism and Resources holds primary responsibility for energy-efficiency policy for the commercial and industrial sectors. It also co-ordinates international liaison on energy-efficiency indicators for the Australian government. Policy development for the household sector falls under the mandate of the Australian Greenhouse Office. The Australian Bureau of Agricultural and Resource Economics (ABARE) undertakes key data analysis for energy indicators and publishes a handbook *(Energy in Australia)* each year that includes energy use information. ABARE has previously published energy intensity data and analytical reports (approximately every three years). ABARE uses the residual-free Log-Mean Divisia Index decomposition method for energy intensity trend analyses.

## Canada

In April 2006, Natural Resources Canada published *Energy Efficiency Trends in Canada*, the 11th edition of Canada's indicator book. This book tracks trends in energy efficiency, energy use and related greenhouse gas emissions over the period 1990 – 2004. The book identifies and analyses the factors that drive changes in energy demand; it also develops indicators to help track improvements in energy efficiency. The book examines efficiency trends across the economy, along with five main end-use sectors: residential, commercial/institutional, industrial, transport and electricity generation. In the latest edition of the book, the Office of Energy Efficiency (OEE) has adopted the residual-free Log-Mean Divisia Index I (LMDI I) decomposition methodology, making a break from the Laspeyres index methodology, which has been used since 1999.

This publication can be downloaded from the following link:
http://oee.nrcan.gc.ca/corporate/statistics/neud/dpa/data_e/publications.cfm?attr=0

## Japan

Japan has used indicators extensively in its Top Runner programme, which seeks continuous improvements in energy conservation standards and performance in the residential, commercial and transport sector. A voluntary action plan uses indicators to set voluntary targets for reducing energy intensity in industry. When calculating energy indicators for industry, careful consideration is given to process boundaries (such as the definition of attached processes), as well as the treatment of sold energy. Close collaboration between government and industry associations is necessary for improved collection, analysis and dissemination of data. Since 1992, the Energy Conservation Center has partnered with the Energy Data Modelling Center of the Japan Institute of Energy Economics to publish the *Handbook of Energy and Economic Statistics in Japan* (an English version has been available since 1996). Covering the period from 1965 to present, this publication provides raw data on energy and economics, highlights key indicators, and analyses many disaggregated energy intensities (travel, freight and household energy uses).

## New Zealand

Energy indicators in New Zealand are supported by the Energy Efficiency and Conservation Authority and provide a mechanism to monitor the National Energy Efficiency and Conservation Strategy, which is reviewed every five years (as required by the *Energy Efficiency and Conversation Act 2000*). A Divisia decomposition is used to separate the effects of activity, structure, quality, weather and technical efficiency. In the past, the indicators were used primarily to track progress. Currently, they are used to evaluate the effectiveness of policy. In the future, indicators will be used to set targets and to select appropriate policies and programmes. In 2000, New Zealand published the second edition of *The Dynamics of Energy Efficiency Trends in New Zealand: A Compendium of Energy End-use Analysis and Statistics*, which covers the period from 1980 to 1998.

## United States

Recognising that many factors – not just efficiency – influence energy services, the US Department of Energy (US DOE) has implemented a new system of energy intensity indicators. These disaggregated indicators separate energy intensity from structural changes, climate and other factors. The system includes several economy-wide indicators, which are further divided into four energy-consuming sectors: transport, industry, commercial and residential. The US DOE uses a hierarchical framework to develop these disaggregated energy indicators at the sector and sub-sector level. Indices are developed for detailed sub-sectors of the economy and then combined to create indicators for more aggregate sectors. The US DOE uses the logarithmic mean Divisia decomposition methodology, with index aggregation across the hierarchy of indicators.

Detailed data on energy intensity trends, dating from 1949, are available on the following site: http://intensityindicators.pnl.gov/

## European Union

The ODYSSEE database on energy-efficiency indicators has become a reference for evaluating and monitoring annual energy-efficiency performances and energy-related $CO_2$ emissions for the EU15 countries and Norway. The database, which includes 200 indicators and covers the period 1970 – 2004, is also used to monitor and evaluate energy-efficiency policies, at both national and EU levels. This database is updated regularly (at least annually) and supports the use of a common methodology to produce comparative indicators for energy efficiency. The data are collected mainly from national sources, with some additional data from Eurostat. Indicators in the ODYSSEE database cover both energy efficiency and $CO_2$. ADEME (France) leads the project with assistance from Enerdata, the technical co-ordinator responsible for collecting submissions from national agencies.

Further information on ODYSSEE can be found at: www.odyssee-indicators.org/

# Austria

The Austrian Energy Agency is the national energy agency and is responsible for the ODYSSEE indicators analysis. Energy statistics are collected and disseminated by Statistik Austria.

# Belgium

Belgium's three regions (Flanders, Wallonia, and Brussels) share responsibility for energy efficiency. A new national climate plan is currently under preparation, which is to be approved by both the federal and regional ministers. The CONCERE/ENOVER committee is responsible for co-ordinating the federal and regional framework for energy-efficiency policies. The Federal Public Service Economy holds the mandate to collect all energy consumption data (with the exception of that from industry) for submission to Eurostat and the IEA. ECONOTEC is responsible for the ODYSSEE indicators analysis.

# Denmark

The Danish Energy Authority determines the national energy balance and is responsible for international reporting. Energy statistics are published annually. Indicators for energy efficiency are prepared annually or biannually, and submitted to the ODYSSEE project. The annual energy balance contains some indicators, but a fuller set of indicators are published annually or bi-annually through ODYSSEE. Time series data on energy and economy are available from 1975 to 2004.

The 2005 version of the Danish Energy Statistics can be found through the following link:
www.ens.dk/graphics/UK_Facts_Figures/Statistics/yearly_statistics/Energy%20St atistics%202005.pdf

# Finland

Energy policy in Finland is set by the Energy Department of the Ministry of Trade and Industry. Motiva Oy is a state-owned company focusing on energy efficiency and renewable energy; it also provides Finland's contribution to ODYSSEE. The energy balance is prepared annually by Statistic Finland.

# France

In France, ADEME is the national agency in charge of implementing energy efficiency, renewable energy sources and environmental policies. It has been the lead agency behind the development of the European Union sponsored ODYSSEE project and now acts as the project's co-ordinator.

# Germany

In Germany, analysis on energy indicators is carried out by the Fraunhofer ISI and disseminated through the ODYSSEE network. In addition to being the German national agency for the ODYSSEE network, the Fraunhofer ISI provides technical support for the ODYSSEE project. Germany's most important source for energy consumption data is the German energy balance, which is supported by the Arbeitsgemeinschaft Energiebilanzen (AGEB) – more specifically, the Working Group on Energy Balances. Energy consumption data from the energy balances at an aggregated level are available up to 2005. The detailed energy balances, which also include sub-sectoral energy consumption (industrial branches, transport modes), are available up to 2002.

# Greece

The Centre for Renewables Energy Source (CRES) was founded in 1989 as the national agency for the promotion and implementation of energy savings, rational use of energy and renewable energy sources. CRES also provides the indicators analysis for Greece to the ODYSSEE network. Greek energy policy is closely aligned with that of the EU.

# Ireland

Established in 2002, Sustainable Energy Ireland (SEI) is Ireland's national energy agency. Its mandate is to promote and assist sustainable energy development. The SEI's Energy Policy Statistical Support Unit is responsible for energy statistics and indicators analysis; it also prepares and submits the ODYSSEE submission. The unit produces an annual publication (*Energy in Ireland 1990 - 2005*, now in its second edition) on energy trends and indicators, as well an annual statistics compendium (*Energy Statistics Publication 1990 - 2005*). Basic intensity indicators are produced for the following sectors: industry, transport, residential and services.

Both publications can be downloaded from the following address:
www.sei.ie/index.asp?locID=686&docID=659

# Italy

The Italian National Agency for New Technologies, Energy and the Environment (ENEA) is responsible for submitting Italy's contribution to ODYSSEE.

# The Netherlands

The Energy Research Center (ECN) of the Netherlands carries out policy studies and provides the Dutch annual contributions to the ODYSSEE project. In 2005, a new Energy Report was published with long-, medium- and short-term goals. Voluntary agreements were set up in 1992 to improve energy efficiency in industry.

# Norway

In Norway, the Institute for Energy Technology (IFE) is responsible for the country's ODYSSEE contribution. ENOVA, a government-owned agency was set up in 2002 to promote energy savings, new renewables and environmentally friendly natural gas solutions. Statistics Norway is responsible for producing energy statistics and is the main source of data for the indicators.

# Portugal

Portugal's Energy Agency, ADENE, contributes to the ODYSSEE network. Portugal's energy balance is prepared by the Portuguese Directorate General of Geology and Energy (DGGE). The DGGE is also responsible for the execution and evaluation of Portugal's energy policy.

# Spain

The Spanish Institute for Diversification and Energy Savings (IDAE) provides technical assistance to the Ministry of Industry, Tourism and Commerce. This national agency is responsible for promoting energy efficiency and renewable energy. The IDAE is responsible for submitting Spain's indicators analysis to ODYSSEE.

# Sweden

The Swedish Energy Agency is responsible for implementing energy policies set by the government and represents Sweden on the ODYSSEE project. It also produces an annual publication on energy indicators, available only in Swedish. *Energy Indicators 2006* presents the use of indicators for monitoring progress toward energy policy targets; it contains five new indicators for the use of oil and 20 updated base indicators. Statistics Sweden compiles energy data statistics, which are published annually in two companion reports: *Energy in Sweden* reviews developments in the Swedish energy sector; *Energy in Sweden: Facts and Figures* provides supporting data.

# United Kingdom

The United Kingdom uses energy indictors as a tool to monitor policy, particularly the country's Kyoto Protocol target, and to analyse trends and make projections. Four key indicators, 28 supporting indicators and more than 100 background indicators are used to monitor the goals of the 2003 *Energy White Paper*. The four key indicators are low carbon, reliability, competitiveness and energy poverty. The 28 supporting indicators explain in more detail the trends shown in the four key indicators. The key and supporting indicators were published in *UK Energy Indicators 2006*, as a supplement to the third annual report of the 2003 *Energy White Paper*. The background indicators are available on line. The 2004 and 2005 editions of the *UK Energy Indicators* included the full range of indicators.

The latest set of indicators are available to download from the following site: www.dti.gov.uk/energy/statistics/publications/indicators/page29741.html

# ANNEX E: ABBREVIATIONS AND GLOSSARY

## Abbreviations

| | |
|---|---|
| $CO_2$ | Carbon dioxide |
| EJ | Exajoules ($10^{18}$ Joules) |
| EU15 | Austria, Belgium, Denmark, Finland, France, Germany, Greece, Ireland, Italy, Luxembourg, the Netherlands, Portugal, Spain, Sweden and the United Kingdom |
| G8 | Group of Eight; member countries are Canada, France, Germany, Italy, Japan, Russia, the United Kingdom and the United States |
| GDP | Gross domestic product |
| GJ | Gigajoules ($10^9$ Joules) |
| GPOA | Gleneagles Plan of Action |
| Gt | Gigatonne ($10^9$ tonnes) |
| HDD | Heating degree-day |
| IEA | International Energy Agency; member countries are Australia, Austria, Belgium, Canada, the Czech Republic, Denmark, Finland, France, Germany, Greece, Hungary, Ireland, Italy, Japan, the Republic of Korea, Luxembourg, the Netherlands, New Zealand, Norway, Portugal, Spain, Sweden, Switzerland, Turkey, the United Kingdom and the United States |
| IEA11 | Australia, Denmark, Finland, France, Germany, Italy, Japan, Norway, Sweden, the United Kingdom and the United States |
| IEA14 | Austria, Canada, Denmark, Finland, France, Germany, Italy, Japan, the Netherlands, New Zealand, Norway, Sweden, the United Kingdom and the United States |
| IEA15 | Austria, Canada, Denmark, Finland, France, Germany, Italy, Japan, the Netherlands, New Zealand, Norway, Spain, Sweden, the United Kingdom and the United States |
| IEA17 | Australia, Austria, Canada, Denmark, Finland, France, Germany, Greece, Ireland, Italy, Japan, the Netherlands, New Zealand, Norway, Sweden, the United Kingdom and the United States |
| IEA19 | Australia, Austria, Belgium, Canada, Denmark, Finland, France, Germany, Greece, Italy, Japan, the Netherlands, New Zealand, Norway, Portugal, Spain, Sweden, the United Kingdom and the United States |
| ISIC | International Standard Industrial Classification of all economic activities, third revision, (ISIC, Rev.3) |

| | |
|---|---|
| kg | Kilogramme |
| kWh | Kilowatt-hour |
| $m^2$ | Square metres |
| MER | Market exchange rates |
| MJ | Megajoules ($10^6$ Joules) |
| Mt | Million tonnes ($10^6$ tonnes) |
| Mtoe | Million tonnes of oil equivalent |
| PJ | Petajoules ($10^{15}$ Joules) |
| pkm | Passenger-kilometres |
| PPP | Purchasing power parities |
| SUV | Sport utility vehicle |
| TFC | Total final consumption |
| TJ | Terajoules ($10^{12}$ Joules) |
| tkm | Tonne-kilometres |
| toe | Tonne of oil equivalent |
| USD | United States dollar |
| VA | Value-added (output approach of measuring GDP) |
| vkm | Vehicle-kilometres |

# Glossary

**Activity** refers to the basic human or economic actions that drive energy use in a particular sector. It is measured as value-added output for manufacturing and services, as population levels in the household sector, as passenger-kilometres for passenger transport, and as tonne-kilometres for freight transport.

**Carbon intensity** is the amount of $CO_2$ emitted per unit of energy use.

**Climate correction** uses heating degree-days to adjust household space heating consumption for year-to-year or country-to-country variations in climate.

**Coal** includes hard coal, lignite/brown coal and derived fuels (including patent fuel, coke oven coke, gas coke, BKB, coke oven gas and blast furnace gas). Peat is also included in this category.

**District heat** is heat distributed from a central heating plant to buildings, factories, etc.

**Decomposition** is the analytical approach used to calculate the effects of various "components" (activity, structure and energy intensities) on aggregate energy use.

**Energy intensity** is the amount of energy used per unit of activity. This publication uses changes in the energy intensity effect as a proxy for developments in energy efficiency.

**Energy services** imply the actual services for which energy is used, e.g. heating a given amount of space to a particular temperature for a period of time. In this study, a quantitative measure of energy service demand in a sector is determined by combining the activity and structure effects.

**Final energy** is the energy supplied to the consumer in each end-use sector, which is ultimately converted into heat, light, motion and other energy services. It does not include transformation and distribution losses.

**Freight transport** includes the domestic haulage of goods by trucks, rail, ships and barges. In this study it does not include air freight transport and pipelines.

**Fuel mix** represents the share of various fuels such as coal, oil, natural gas, heat and electricity that make up final energy use.

**Gasoline equivalent litre** is a concept used to compare the energy content of gasoline and other road transport fuels (e.g. diesel) which have different calorific values than gasoline. In order to properly aggregate physical quanities of different fuels they are expressed in quantities energetically equivalent to a litre of gasoline.

**Gross domestic product (GDP)** is a measure of economic activity, defined as the market value of all final goods and services produced within a country (output approach). In this publication, GDP figures are given for calendar years, expressed in 2000 USD. The conversion from national currency to USD is done using either purchasing power parities or market exchange rates.

**Heating degree-days** are calculated from the difference between the average daily outdoor temperatures and a reference temperature (18° Celsius or 65° Fahrenheit). They are used in this publication to provide climate corrections for energy use in household space heating.

**Households** cover all energy-using activities in apartments and houses, including space and water heating, cooking, lighting and the use of appliances. It does not include personal transport.

**Manufacturing** covers finished goods and products for use by other businesses, for sale to domestic consumers, or for export. Total manufacturing is divided into the following key industries: food, beverages and tobacco; paper, pulp and printing; chemicals; non-metallic minerals; primary metals; metal products and equipment; and other manufacturing. The fuel-processing industries and fuels used as feedstocks are not included.

**Natural gas** includes gas works gas but excludes natural gas liquids.

**Oil** comprises crude oil, natural gas liquids and petroleum products, such as heavy fuel oil, gas/diesel oil, liquefied petroleum gas, motor gasoline and kerosene.

**Other fuels** includes geothermal, solar, wind, tide and wave energy. The "Other" category is very small in the end-use sector.

**Passenger-kilometres** are a measure of transport activity and are calculated by multiplying the number of kilometres a vehicle travels by the number of passengers. For example, if a vehicle carries two passengers for one kilometre then it has travelled two passenger-kilometres (but only one vehicle-kilometre).

**Passenger transport** includes the movement of people by road, rail, sea and air. Road transport is sub-divided further into cars and buses. In this study, only domestic air travel is included; international air travel is not covered.

**Renewables** comprise biomass and animal products (wood, vegetal waste, ethanol, animal materials/wastes, etc.), municipal waste and industrial waste.

**Savings** refer to the difference between the hypothetical energy use (or hypothetical $CO_2$ emissions) and actual energy use (or actual $CO_2$ emissions).

**Services** include activities related to trade, finance, real estate, public administration, health, education and commercial services.

**Structure** represents the mix of activities within a sector, e.g. shares of each sub-sector in manufacturing, energy end-uses in households, or the modal mix in passenger and freight transport.

**Tonne-kilometres** are a measure of freight transport activity. For example, if a truck carries a load of two tonnes for one kilometre then it has travelled two tonne-kilometres (but only one vehicle-kilometre).

**Useful energy** is calculated as final energy minus losses estimated for boilers, furnaces, water heaters and other equipment in buildings. It is used for estimates of heat provided in space and water heating.

**Vehicle-kilometres** are a measure of transport activity. For example, if a vehicle travels one kilometre per day for a year, then it will have travelled a total of 365 vehicle-kilometres (see also passenger-kilometres).

# ANNEX F: REFERENCES AND FURTHER READING

## References

IEA (2004), *Oil Crises and Climate Challenges: 30 Years of Energy Use in IEA Countries*, IEA/OECD, Paris.
www.iea.org/Textbase/publications/free_new_Desc.asp?PUBS_ID=1260

IEA (2006a), *Energy Technology Perspectives: Scenarios & Strategies to 2050*, IEA/OECD, Paris.
www.iea.org/w/bookshop/add.aspx?id=255

IEA (2006b), *World Energy Outlook 2006*, IEA/OECD, Paris.
www.worldenergyoutlook.org/

IEA (2007), *Tracking Industrial Energy Efficiency and $CO_2$ Emissions*, IEA/OECD, Paris.
www.iea.org/w/bookshop/add.aspx?id=298

## Further Reading

ADEME (2005), *Energy-efficiency Monitoring in the EU-15*, ADEME, Paris.
www.odyssee-indicators.org/Publication/chapters.php

IAEA (2005), *Energy Indicators for Sustainable Development: Guidelines & Methodologies*, International Atomic Energy Agency, Vienna.
www-pub.iaea.org/MTCD/publications/PDF/Pub1222_web.pdf

IEA (2005), *Energy Statistics Manual*, IEA/OECD, Paris.
www.iea.org/Textbase/publications/free_new_Desc.asp?PUBS_ID=1461

IEA (2006), *Light's Labours Lost – Policies for Energy-efficient Lighting*, IEA/OECD, Paris.
www.iea.org/w/bookshop/add.aspx?id=302

IEA PUBLICATIONS, 9, rue de la Fédération, 75739 PARIS CEDEX 15
PRINTED IN FRANCE BY STEDI MEDIA
(61 2007 24 1P1) ISBN : 978-92-64-03429-7 – SEPTEMBER 2007

# The Online Bookshop

International Energy Agency

All IEA publications may be bought
online on the IEA website:

**www.iea.org/books**

You may also obtain PDFs of
all IEA books at 20% discount.

Books published before January 2006
- with the exception of the statistics publications -
can be downloaded in PDF, free of charge,
from the IEA website.

## IEA BOOKS

*Tel: +33 (0)1 40 57 66 90*
*Fax: +33 (0)1 40 57 67 75*
*E-mail: books@iea.org*

International Energy Agency
9, rue de la Fédération
75739 Paris Cedex 15, France

---

**CUSTOMERS IN
NORTH AMERICA**

**Turpin Distribution**
The Bleachery
143 West Street, New Milford
Connecticut 06776, USA
Toll free: +1 (800) 456 6323
Fax: +1 (860) 350 0039
oecdna@turpin-distribution.com

www.turpin-distribution.com

*You may also send*

*your order*

*to your nearest*

*OECD sales point*

*or use*

*the OECD online*

*services:*

**www.oecdbookshop.org**

**CUSTOMERS IN
THE REST OF THE WORLD**

**Turpin Distribution Services Ltd**
Stratton Business Park,
Pegasus Drive, Biggleswade,
Bedfordshire SG18 8QB, UK
Tel.: +44 (0) 1767 604960
Fax: +44 (0) 1767 604640
oecdrow@turpin-distribution.com

www.turpin-distribution.com